W9-BFX-412

Stargazer

The Life and Times of the Telescope

Fred Watson

DA CAPO PRESS

A Member of the Perseus Books Group

For Trish

Set in 11/14.5 point Minion by Midland Typesetters

Cataloging-in-Publication data for this book is available from the Library of Congress.

First Da Capo Press edition 2005
Originally published in 2004 by Allen & Unwin; reprinted by arrangement
ISBN 0-306-81432-3

Published by Da Capo Press
A Member of the Perseus Books Group
www.dacapopress.com

Da Capo Press books are available at special discounts for bulk purchases in the U.S. by corporations, institutions, and other organizations. For more information, please contact the Special Markets Department at the Perseus Books Group, 11 Cambridge Center, Cambridge, MA 02142, or call (800) 255-1514 or (617) 252-5298, or e-mail special.markets@perseusbooks.com.

1 2 3 4 5 6 7 8 9—08 07 06 05

CONTENTS

FOREWORD

The story of the telescope goes back now for almost four hundred years. It is fascinating for many reasons, not least because of the extraordinary personalities who have been involved. It is also staggering to realise the progress that has been made since the time when Galileo's tiny 'optick tube' was turned toward the sky. Today we have telescopes powerful enough to peer into the far reaches of the Universe.

Fred Watson is uniquely qualified to write this book. He is one of the world's leading astronomers, and he has been personally responsible for some very important developments—notably with regard to optical fibres, which now play a vital part in modern astronomical research. Almost equally important in the present context is that he is a skilled writer of books for non-specialists, and has the ability to make difficult problems seem easy. He is also a 'natural' as a broadcaster on radio and television—as I well know since we have broadcast together on many occasions.

He has a marvellous story to tell, and he has done so in his unique way. This is a book which will be enjoyed by beginners and specialists alike—also by readers who do not pretend to have any particular interest in the sky. I am proud to have been asked to write the Foreword.

Patrick Moore
Selsey

ACKNOWLEDGEMENTS

'This damn book nearly drove me mad.' So wrote the late Spike Milligan in the foreword to his first novel, and I know exactly how he felt. More to the point, *Stargazer* nearly drove my family mad, and it is to them I owe the biggest 'thank you'. Without the unflagging support of Trish, James, Will and (by remote control from Scotland) Helen and Anna, the task would have been impossible.

It would also have been impossible without the enthusiastic assistance of Sandra Ricketts, librarian of the Anglo-Australian Observatory in Sydney. Sandra did her utmost to find every obscure reference I asked for, and then, at extremely short notice, took on the daunting task of picture research without batting an eyelid. Karen Moran, librarian of the Royal Observatory Edinburgh, cheerfully guided me through the Crawford Collection. A number of other librarians were very helpful, most notably at Queen's University, Belfast and the University of Western Sydney. Paul Cass of the Anglo-Australian Observatory generously gave me access to his private collection of astronomy books.

I thank Sir Patrick Moore not only for his kind words in the Foreword, but also for a lifetime of inspiration. In a similar vein, David Malin has provided far more than just colour pictures in this book. And to my colleagues at the Anglo-

Australian Observatory, both at Siding Spring and in Sydney, I say a big thank you for your support.

Stargazer owes a lot to the help and encouragement of other colleagues over the years, and I thank Peter Abrahams, Sukdeep Aulakh, Richard Bingham, Brian Boyle, Marilyn Campbell, Matthew Colless, Roger Davis, John Dawe, Hayden Gabriel, Ben Gascoigne, Peter Gillingham, Ian Glass, Tom Jarrett, Kevin Johnson, Nick Lomb, Alison Morrison-Low, Iain Nicolson, Paddy Oates, Wayne Orchiston, John Perdrix, Gilbert Satterthwaite, Allen Simpson, David Sinden, John Watson and Sue Worswick. I also acknowledge the debt of gratitude I owe to my first boss, the late David Brown.

It was a perceptive suggestion by Ian Bowring of Allen & Unwin that led to the book, and his guidance and continuing enthusiasm during its long gestation period have been invaluable. It has also been a privilege to work with the company's senior editor, Emma Cotter, in whose truly capable hands the book has taken shape. Thanks, too, to Marie Baird, Elizabeth Bray, Jo Paul, Emma Singer and Ruth Williams of Allen & Unwin.

Stargazer draws heavily on the published research of Jim Bennett, Ronald S. Brashear, Randall C. Brooks, Allan Chapman, John R. Christianson, Owen Gingerich, Richard F. Harrison, Raymond and Roslynn Haynes, John B. Hearnshaw, Norriss S. Hetherington, Alan W. Hirshfeld, Michael Hoskin, David Leverington, Anita McConnell, Richard McGee, the late Colin A. Ronan, Engel Sluiter, Stuart Talbot, the late Victor E. Thoren, Albert Van Helden, Brian Warner and Ray N. Wilson. To these accomplished historians of astronomy, I express my admiration and gratitude.

It is a pleasure to thank Hélène and Henrik Wachtmeister, owners of Tycho Brahe's birthplace at Knutstorp, for their generous hospitality when I visited in 1996. Finally, to Angela

Catterns, Ross and Helen Edwards, Malc and Laura Hartley, Jen Lacey, James O'Loghlin and William and Nina Reid— thank you for the inspiration!

PICTURE CREDITS

The following institutions and individuals have kindly provided the images contained in these pages, and their help is gratefully acknowledged.

Crawford Collection, Royal Observatory Edinburgh (pp. 22, 24, 26, 30, 94, 97, 149)

States General Manuscripts via Albert Van Helden 'The invention of the telescope' (p. 61)

Petworth Manuscripts via *Journal for the History of Astronomy* (p. 72)

Thomas Seminary, Strasbourg (p. 78)

US Naval Observatory Library (pp. 102, 104)

Dibner Collection (p. 119)

Royal Society (pp. 126, 129)

Royal Astronomical Society via Ray Wilson *Reflecting Telescope Optics I* (p. 158)

Padua Observatory (p. 162)

Royal Astronomical Society via *Journal for the History of Astronomy* (p. 166)

Anglo-Australian Observatory Library (pp. 171, 192, 206, 212, 214, 236, 243)

Physical Sciences Library, University of New South Wales (p. 183)

Tartu Observatory (p. 188)

Kensington and Chelsea Public Libraries via *Journal for the History of Astronomy* (p. 196)

Master and Fellows of Trinity College Cambridge, courtesy Ian Glass (p. 199)

David Sinden (p. 201)

Strand Magazine, 1896, courtesy Ian Glass (p. 221)

Australian National University (p. 228)

The Engineer, 1886, courtesy Ian Glass (p. 245)

Carnegie Institute of Washington (p. 254)

Australia Telescope National Facility Library (p. 274)

PROLOGUE

In one of his acclaimed *Far Side* cartoons, Gary Larson depicts three astronomers working at a giant telescope. One sits at the eyepiece while the others snigger gleefully behind his back. A thick black ring has appeared around his eye, and a barely concealed marker-pen leaves little doubt as to how it got there. It's a good cartoon and, like all Larson's quirky, off-beat snapshots, it raises a smile.

To the real-life working astronomer, it brings an added delight. Whether or not it's intentional, the stereotyping is perfect. Larson's cartoon has all the clichés. His astronomers are male, middle-aged and nerdish in their white lab coats. The telescope is a hugely bloated version of an old-fashioned extending spyglass, poking out of its dome into the starry night beyond. And you can bet your life that if you could see the other end, you would find a giant lens framing a grotesquely magnified cartoon eye—encircled, of course, with a black ring.

This popular conception of the way astronomers work is wrong in almost every respect. You only have to look at the young men and women currently completing their doctoral studies in astronomy. For the most part they are the antithesis of nerds—they are poised, gifted and highly personable. And they would be the first to tell you that even though today's great telescopes still bear the name chosen by a Greek poet almost four centuries ago, they look nothing like their forebears. They are high-tech, space-frame structures containing not giant lenses but giant mirrors, perfectly curved into shallow dishes to capture and focus the incoming light. They don't protrude from their enclosures. And no one ever—*ever*—looks through them.

Perhaps this last aspect is the one that surprises and disappoints most people. The thought of a telescope feeding starlight straight into a piece of intelligent hardware to be coldly analysed and measured—and then written in digital form to a network of computers—de-romanticises the whole idea of what astronomers do. But human eyes compare very unfavourably with sensitive TV-type detectors. Computers are infinitely better at recording tedious numerical data than people. And, as in all science, it is the objective gathering of hard facts that allows our understanding to progress, and theories to be built and tested.

In astronomy, these theories relate to some of the most profound questions facing humanity. Where did we come from? Are we alone in the Universe? What is our destiny? De-romanticised or not, mystery and intrigue abound in the ultimate objectives of the science.

There is one aspect of the *Far Side* cartoon that does capture reality rather well. For although they seldom resort to practical jokes (and when they do, they're usually *very* funny), astronomers are intensely competitive individuals. They are driven not only by the desire to reveal new insights into the mysteries of the Cosmos, but also to be the first to do so. Being an also-ran is not an option. And, in consequence of that, they are very competitive about the telescopes they use. They want them to be the best available. They always have.

In a reversal of the trend towards miniaturisation in everything technological, telescopes improve as they get bigger. The end result is that astronomers tend to be very concerned with size when they plan their observations. And this can develop into an unusual kind of megalomania known colloquially as 'aperture fever'.

The term 'aperture' simply means the diameter of a telescope's light-collecting surface—its mirror. Victims of aperture fever are obsessed with the idea that nothing else is important. While this term originated in the world of amateur astronomy, where enthusiasts long for ever-bigger telescopes with which to view the heavens from their backyards, the disease afflicts professional astronomers, too.

Paradoxically, the fever is simultaneously the driving force of telescope-building and its scourge. Without it, telescopes would never have progressed to the scale of today's giants. The history of the telescope—certainly over the last two centuries—has been dominated by single-minded individuals whose motivation was to build a bigger telescope than the last one. The problem arises when the last one is discarded like yesterday's newspaper, closed down prematurely before it has reached its full potential in bringing home new discoveries. Such short-sightedness discounts the fact that telescopes have inherently long working lives.

In a sense, this book is a history of aperture fever—though it would be wrong to expect a bizarre romp through the outer fringes of human psychology. Rather, it tells the story of the telescope-makers and the instruments they produced. It is as much about triumph and tragedy as steelwork and glassware. And it is an ongoing story, for both the heartache and the technological development continue today. Aperture fever is endemic, and its consequences are there for all to see in the world's growing array of giant telescopes.

1

POWER TELESCOPES

BOLDLY INTO THE NEW MILLENNIUM

There is no better way to sample the state of the art of telescope-building—nor the promise of what is to come—than by attending an international telescope symposium, and that is where we start our story. Such events are not everyday occurrences, but the year 2000 saw a gathering of such significance that its deliberations will reverberate long and loud through the annals of astronomy. It was a very large meeting, embracing no less than thirteen separate conferences. It attracted 1300 scientists, engineers, directors of institutions and household-name professors, all with a common interest in the tools of the astronomer's trade. In that broad forum, the mysteries of the Universe met the nuts and bolts of engineering, and in the willing hands of its participants lay the future of the telescope.

The symposium's title brashly declared its spirit: 'Power Telescopes and Instrumentation into the New Millennium'. It took place in Munich, during the last week in March—

although the message that this was supposed to be springtime obviously hadn't got through. By turns, the participants endured icy winds, rain, sleet and snow on their daily journey to the symposium sessions. Only on the last day did the Sun finally put in an appearance.

In some ways, the symposium inside Munich's International Congress Centre was as tumultuous as the weather outside—although its cut and thrust were veiled in the subdued tones of academia. Let us join the participants, and eavesdrop on their deliberations. But, before we do, we should explore their passions a little further.

First, there was chauvinism. A rare species of chauvinism, to be sure—but that's what it was. It concerned the different varieties of telescope astronomers use today.

Until 1932, when Karl Jansky of the Bell Telephone Laboratories discovered radio waves coming from space, there was only one kind of telescope. It collected and focused ordinary visible light. It used the science of optics to allow information to be retrieved—at first by human eyes, and then, from the 1880s, by photographic plates.

Today, there are as many different kinds of telescope as there are varieties of natural radiation traversing the Universe. A plethora of names identifies these ghostly emissions— gamma rays, X-rays, ultraviolet rays, visible light, infra-red, millimetre waves (microwaves) and radio waves. Arranged in order of wavelength, they form the electromagnetic spectrum. Somewhere near the middle is the radiation to which our eyes are sensitive. Its wavelength is measured in nanometres (millionths of a millimetre), and ranges from about 400 nanometres (nm) for violet light to about 700 nm for deep red. In between lies the rainbow of the visible spectrum.

One of the heartwarming facts astronomers have brought to us in recent years is that the Earth is constantly bathed in radiation covering the whole electromagnetic spectrum. It comes from sources everywhere in the Universe. Much of it never reaches the surface of the planet because it is absorbed by the atmosphere. If you want to observe X-rays or gamma rays, for example, you have to mount specialised telescopes on spacecraft. But for some categories of radio and infra-red radiation—and visible light, too, of course—the observations can be made from the ground.

To distinguish them from their more exotic cousins, telescopes that use visible light are now called optical telescopes. They require darkness to operate, so observing with them is always night work. They also require clear skies. And, despite the effectiveness of the glamorous new flavours of astronomy, optical telescopes still play a vital role in the study of the Universe. It is the central position of visible light in the electromagnetic spectrum—and the fact that ordinary stars emit most of their energy as visible light—that keeps them in such high demand.

Of course, optical telescopes also predate all the other types by more than 300 years. So perhaps it was not surprising that those attending our forum paid little more than lip-service to the newer, 'invisible' astronomies. There was a clear underlying message: at this meeting, optical astronomy ruled, OK. Wavelength chauvinism was alive and well in Munich.

Then there was the obsession with size. Just *why* do optical telescopes have to be so big?

Unlike Gary Larson's cartoon telescope, today's real telescopes have at their heart a shallow, concave mirror to collect and focus the incoming light. Bigger mirrors collect and concentrate more light, and an insatiable appetite for light—even in very small quantities—is the most common dietary complaint among astronomers. The more light that can be collected, the fainter the objects that can be studied.

But there is another craving that draws astronomers towards ever bigger telescope mirrors, one known as resolution—the fineness of the detail that can be seen in a magnified image of the sky. The quest for resolution is as old as the telescope itself, for it was the instrument's ability to reveal invisible detail that made it such an astonishing invention in the first place. Today, the physics of the situation is well understood: given the necessary degree of optical perfection, the bigger the telescope mirror the finer the detail it is capable of recording.

Like all dimensions in the sky, resolution is measured as an angle. It is expressed in arcseconds—microscopic units that are to angles what nanometres are to length. Geometry tells us that an arcsecond is 1/3600th of a degree. So much for geometry; it's much more instructive to imagine a person 5 kilometres away holding up a coin. An Australian dollar, a British pound and a US quarter are all about the right size. To your eye, the coin's diameter at that distance is one arcsecond—and you would need a sizeable telescope to be able to see it.

Putting some figures on resolution, a one-metre diameter telescope mirror is theoretically capable of showing detail on a scale of a little more than 0.1 arcseconds—the coin at 50 kilometres. But a 4 metre mirror could resolve detail of one-quarter the size—0.03 arcseconds. That is fine enough to detect surface markings on the planet Pluto, or the disc of the giant star Betelgeuse. Bigger is definitely better.

Unfortunately, there is a wholly unwelcome natural phenomenon that plays havoc with resolution, and that is atmospheric turbulence. We're all familiar with what happens when a jet aircraft ploughs into turbulent air 10 kilometres or so above the ground. The nerve-racking shaking and juddering happens even in cloudless skies. That same turbulence has an equally alarming effect on rays of light coming down through the atmosphere. It gives the stars their appealing twinkle when seen with the naked eye—but in the telescope, what should be infinitesimally small points of light are blown up into fuzzy, trembling balls.

For more than three centuries, astronomers have described the degree of turbulence in the atmosphere as 'seeing', and the term now refers specifically to the diameter of the diffuse star-images it produces. The very best seeing in perfectly stable air might yield star-images as small as 0.3 arsceconds in diameter; poor seeing can blow them up to 3 arsceconds or more. Either way, the exquisite detail that the telescope is capable of recording is lost altogether among inflated blobs of light.

SEEING THE UNSEEABLE

What can be done about the problem of seeing? The direct approach is to take your telescope above the atmosphere, but that is *very* expensive. The Hubble Space Telescope, launched in 1990, was designed primarily with this in mind (though its high-level vantage point also provided unprecedented access to the ultraviolet waveband). The story of the Hubble, its flawed 2.4 metre mirror and the 1993 rescue mission that enabled engineers to recover most of its intrinsic resolution is well known, but less widely appreciated is its cost. The eventual bill to build, launch and fix it was well over US$2 billion (1990 dollars), and by the time the project is completed sometime beyond 2010, it will have notched up more than US$6 billion.

Such sums are almost unheard of in the scientific world, except for a *very* few big-ticket items. Eventually, the Hubble will be succeeded by a larger instrument called the James Webb Space Telescope (JWST), which will be launched in 2011. That instrument will be very different from its predecessor. It will have a mirror 6.5 metres in diameter, and will work in the infra-red. Moreover, it will ply its trade from a point some 1.5 million kilometres from Earth rather than in low Earth orbit. Nevertheless, its optimistic proponents expect it to be much cheaper than the Hubble.

Both Hubble and JWST are specialised telescopes, and astronomers have always recognised that their wider needs can only be met by building less expensive ground-based instruments. The Hubble's US$2 billion price-tag, for example, would be enough to build *twenty* of today's largest class of ground-based telescopes. So astronomers have had no choice but to confront atmospheric turbulence head-on.

Some places on Earth tend to suffer more from bad seeing than others. During the 1960s, scientists all over the world embarked on site testing programmes to establish where the best observing conditions could be found, spurred on by a new generation of wide-bodied jets that promised easy access to remote facilities. Before this, telescopes had been built wherever the astronomers happened to be—which was not always a good spot.

A few places emerged with that rare combination of clear skies, freedom from artificial light and good seeing that is astronomy's Holy Grail. Geography played a vital part. Typically, these places were in middle latitudes (between 20 and 40 degrees north or south of the equator), on mountaintops higher than about 3500 metres (11 500 feet), and near the eastern boundary of an ocean. If the mountain peak was on an offshore island and streamlined with respect to the prevailing wind, so much the better.

In the northern hemisphere, such sites were found in the south-western USA, the Big Island of Hawaii and the island of La Palma in the Canaries. Continental Europe was largely ruled out by its bright lights and indifferent weather. In the south, the peaks of northern Chile and the high Karoo of southern Africa were favoured. Australia, having no high mountains on its western seaboard, could boast no truly excellent sites—although the Siding Spring Observatory in central New South Wales is still one of the world's least polluted by artificial light. In recent years, a few other places have shown promise. Antarctica, for example, boasts conditions on the high plateau near the South Pole that are particularly good for observing in the infra-red—even though night only darkens the continent for half the year.

When large telescopes were built on high mountain sites, it was found that the typical seeing was 0.5 to 1.0 arcseconds— good, but still nowhere near the inherent resolution of the telescopes. Then a promising new technology emerged. It prompted a revolution in the esoteric field of astronomical instrumentation.

By 'instrumentation', astronomers usually mean the auxiliary devices that are attached to their telescopes. In many ways, they are the real tools of the astronomer—the telescopes are merely there to deliver light to them. They range from ultrasensitive electronic cameras to spectrographs—magical devices that sift the rainbow spectra of celestial objects wavelength by wavelength to reveal vital statistics across the abyss of space.

But during the late 1980s, instrument designers began to promise the impossible. They proposed devices that would cancel out the blurring effects of the Earth's atmosphere, allowing objects to be investigated in unprecedented detail. This celestial conjuring trick would be performed using declassified Star Wars technology known as adaptive optics. Small, deformable mirrors, bending in sympathy with the distortions in the incoming light, would simply cancel out the blurring. Early experiments proved so successful that astronomers began to envisage a time in the not-too-distant future when ground-based telescopes might be limited only by their intrinsic resolution. At last, the curse of atmospheric seeing would be broken. And the participants in 'Power Telescopes and Instrumentation into the New Millennium' were eager to hear the latest on the progress of adaptive optics.

POWER TELESCOPES

These three things—a fixation with ground-based optical telescopes, an obsession with size and a feverish excitement at the prospects for adaptive optics—were the basic ingredients that fermented together in the warm lecture theatres of Munich's International Congress Centre.

Placing them against the backdrop of the telescope's progress over the previous hundred years, it is easy to see why they made such a potent brew. At the turn of the twentieth century, a telescope with a mirror 1.5 metres (60 inches) in diameter was considered large. In 1918 came a 2.5 metre (100 inch) telescope and, by 1948, a 5 metre (200 inch) was ready for work. But of greater significance than these individual achievements was the proliferation of 4-metre class telescopes during the 1970s and early 1980s. Eight of them around the world provided flagship tools for a new generation of astronomers. They revolutionised our understanding of the Universe, expanding its known horizons by billions of light-years and populating it with exotic new objects: neutron stars, black holes and quasars.

In Munich, though, new benchmarks were being laid down. Technology had moved on since the 1970s, and engineers had learned how to make bigger telescopes. Then, a state-of-the-art mirror for a 4-metre class telescope had been a thick, inert slab of glass-ceramic material that relied on sheer bulk to maintain the profile of its reflective front surface. But now computer-controlled mechanical fingers could hold much thinner mirrors in shape, no matter what angle the telescope was pointing at. And the mirrors didn't have to be made of a single piece of glass; they could be segmented from many smaller, hexagonal pieces held in precise alignment by the same sort of intelligent control.

With lighter mirrors came lighter telescope structures and a spidery appearance that belied their inherent rigidity. Enclosure design became vastly more sophisticated than the simple domes used previously, allowing air to flow freely and smoothly past the telescope without causing local turbulence. And improved control systems promised telescopes that could be pointed at their targets with sub-arcsecond accuracy. With all these developments, the stage was set for a repeat performance of the 1970s—only this time with 8-metre class telescopes.

An 8 metre mirror is huge; it is the size of a suburban backyard. Its reflective surface has to be of such sublime perfection that if the mirror could be magnified to the size of the Earth, the biggest irregularity would be no higher than a doorstep. That is one reason why the price of an 8-metre class telescope approaches US$100 million, and why multinational consortia have to be formed to build them.

In the decade 1994–2004, no less than ten ground-based telescopes with 8 to 10 metre mirrors would be completed. One of these instruments actually contained two 8 metre mirrors; another incorporated four. Three more telescopes with 6.5 metre mirrors were under construction. This was a veritable explosion of telescope-building. And the participants in the Munich symposium sensed, with some justification, that they were at its epicentre. The meeting highlighted the world's burgeoning suite of 8-metre class telescopes and the leap in light-collecting area and focusing power they presented to astronomers: power to observe celestial objects fainter and more remote than ever before—objects 'close to the edge of the Universe', as the tabloid newspapers were wont to put it.

But therein lay the controversy. For if these new telescopes were to be so effective in unlocking the secrets of the sky, what was the point in maintaining the old ones? Shouldn't the 4-metre class telescopes be consigned to museums in the face of the new machines? Or, at least, demoted to second-rank instruments with minimal support—particularly those on poor observing sites . . . ?

It has to be said that it was not so much the directors and project managers of the new facilities who were responsible for fostering this view. Nor was it the organisers of the symposium. But aperture fever stalked the corridors and halls of the International Congress Centre like a prowling tiger, and voices could be heard urging radical cuts in the funding of telescopes that had until recently been among the world's largest.

The organisers of the symposium had clearly attempted to

defuse the issue through the way they had programmed the meetings. Their view was that there is still so much to be learned about the Universe, both near and far, that *every* telescope pointing to the sky has a valid and useful part to play. As long as the funding can stretch that far, they should all be utilised. And what made that view so evident was the prominence they had given to the symposium's other major attraction—the presentations on auxiliary instrumentation.

The field of astronomical instrumentation is one that rewards clever and innovative thinking without incurring the enormous costs of new telescopes. Typically, it involves dazzling technology of the kind used in adaptive optics. But it extends far beyond that limited area. For example, imagine a spectrograph that can compensate for poor atmospheric conditions by allowing not one but hundreds of objects to be observed at a time. It would allow a 4-metre class telescope on an indifferent site to explore novel and quite unique niches of astronomical research. That is exactly the approach that had been taken during the 1990s with the 3.9 metre Anglo-Australian Telescope at Siding Spring Observatory, one of the oldest of the 4-metre class machines, dating from 1974.

Innovative instrumentation can revitalise old-fashioned telescopes and transform them into giants of astronomical productivity. It is the real equaliser in the balance of telescope superiority—and the true antidote of aperture fever. 'Power Telescopes and Instrumentation into the New Millennium' boasted no fewer than 160 presentations on auxiliary instruments for ground-based telescopes, and another 100 on adaptive optics. One might have thought that this kind of prominence would have averted an aperture fever epidemic altogether.

And indeed, it almost did.

STARVING THE FEVER

Half a world away, on a mountain peak in Chile called Cerro Paranal, the world's biggest optical telescope was in its final stages of assembly. This giant instrument, being built by the European Southern Observatory, incorporated four separate 8-metre class telescopes that could be used independently or linked together to mimic a single 16 metre dish. Three of them were already in operation. For all the linguistic elegance of the European partnership that had given it life, it sported a very ordinary name. It was, and remains, the VLT—the Very Large Telescope.

Even as the VLT was being built, there was talk of still larger telescopes. At the last European conference on optical telescopes, four years earlier, a handful of proposals for amazing instruments with much larger mirrors had been presented. One had been called the ELT—the Extremely Large Telescope—and the name had stuck as a generic term for a new telescope class: ones with mirrors 25 metres in diameter. Mirrors for these giants would not be made from single slabs of glass but from assemblies of smaller pieces under computer control—the now-proven segmented-mirror technology.

The thinking that had raised eyebrows in 1996 had become commonplace by 2000—'from wild to mild', as one participant put it. It had brought with it a whole new vocabulary of telescope names. For example, Caltech offered CELT (the California Extremely Large Telescope), while a Swedish university consortium championed SELT (the Swedish Extremely Large Telescope—which has since been renamed Euro50). On the other hand, the somewhat pretentious MAXAT (the Maximum Aperture Telescope) was about to be discarded in favour of the GSMT (the Giant Segmented-Mirror Telescope) by its own enthusiastic proponents. Enthusiasm notwithstanding, none of these proposals had got remotely near the construction stage and only CELT had any real prospect of being funded in the near future.

But on the final day of the Munich symposium, in a session entitled 'Extremely Large Telescopes', a capacity audience was stunned to learn just how far this line of reasoning had progressed. The logic went like this: if segmented-mirror technology would allow a 25 metre telescope to be built, why not a 50 metre one? Or even a 100 metre one? Surely the remaining engineering challenges were just . . . engineering? Wasn't it merely a question of adding more segments to the mirror and building a bigger structure to support it and point it in the right direction? So, with the kind of pizzazz normally reserved for the launch of next year's BMW sedan, a project called OWL was ceremoniously launched into the astronomical world.

OWL was to be a telescope of no less than 100 metres aperture with a mirror made up of hexagonal segments measuring 2.3 metres across—a staggering 1600 of them. They would be carried and pointed by a structure weighing 14 000 tonnes, and the entire assembly would be in the open air, with a sliding hangar to protect it when not in use. OWL's sharp-eyed vision would, of course, require the elimination of the blurring caused by atmospheric turbulence. It would use a bizarre new technique called multi-conjugate adaptive optics, which involved firing multiple lasers into the upper atmosphere to create a constellation of artificial stars for sensors to lock on to. In that way, the exquisite resolution of the mirror could be fully recovered. The telescope would be able to see detail in the sky on a scale of 0.001 arcseconds—a milli-arcsecond. Such unprecedented resolution would give OWL astonishing capabilities in terms of the celestial objects it would be able to detect. Most of the visible Universe—literally—would fall within its grasp.

The promoters of OWL anticipated that it could be built within twelve years, and be delivering front-rank science five years later. All for a cost of US$1 billion. And the meaning of the acronym? Overwhelmingly Large. What else?

Overwhelming was exactly the effect the presentation had on the audience. The prospect of a telescope with more than a hundred times the light-collecting area of an 8-metre for only ten times the cost sounded like the bargain of the millennium, and it was greeted with rapture. And the prospect of milli-arcsecond resolution simply blew the participants away. Even though OWL was as yet totally unfunded, aperture fever spread through the symposium like wildfire. Within half an hour, it had reached plague proportions.

Curiously, OWL's principal advocates appeared ambivalent about the project. One of them wryly made the point that perhaps the instrument should really be called EGO—the Extra-Giant Optical telescope. And maybe it would eventually turn out to be the ULT—the Unnecessarily Large Telescope. Could the fact that these people were also responsible for the VLT have prompted their caution? One day, even their own Very Large Telescope might fall victim to aperture fever.

The epidemic that followed OWL's spectacular entry into the symposium only served to widen the chasm between the opposing factions of moderation and megalomania. Scepticism about the practicalities of building OWL became palpable among those who remained uninfected by the fever. The multi-conjugate adaptive optics system was seized upon as a potential show-stopper, since high resolution was vital to the scientific viability of the project. Everyone agreed that if adaptive optics didn't work and you were going to be limited by atmospheric seeing, the telescope really wasn't worth building.

And some participants simply remained obstinately unimpressed. One insisted on talking only of the NBT—the Next Big Thing—whatever it might be. Another was even more forthright. Coming late into a glitzy presentation on OWL, this eminent scientist stood at the back of the hall and listened for a few moments. Then, with finality, he rudely broke wind and walked out again.

As the last day of the symposium wore on, participants began to take stock. They had witnessed for themselves the state of the art of telescope-making at the turn of a new millennium. They had seen a vision of the future, with a football stadium-sized telescope harvesting the secrets of the Universe, and perhaps discovering everything there is to be known. But what they had not been given was any sense of how the telescope had found its way to this point in its development—how, as it approached its 400th birthday, it had been shaped by the dreams and aspirations of earlier generations of astronomers and sculpted by the technological realities they faced. History had played no part in these proceedings. And, while that was entirely understandable in such a futuristic, space-age gathering, it had left the symposium with an air of clinical soullessness.

It was history that finally restored a sense of proportion to the symposium and put humanity back on to the agenda. But it was history of a different kind: modern-day history, intertwined with recent political turmoil.

The last presentation in the adaptive optics conference—almost the last of the whole symposium—was given by two professors from a former communist bloc country. In faltering English, one slowly read a prepared text while the other illustrated it with old-fashioned overhead projector slides—a dramatic contrast to the slick presentations most speakers had conjured up from their laptop computers. The subject matter was of the highest quality, describing plans for adaptive optics on a proposed 25 metre telescope in the ELT class. But when, at the end of what had clearly been an ordeal, they were questioned about the time-scale on which this telescope would be built, they simply smiled and shrugged. 'We don't know,' they said. 'There are no prospects that it will ever be funded.'

Within the previous year, perhaps half the audience would have received electronic mail from academic colleagues in this same eastern bloc nation asking for help with the barest necessities of life. Food, clothing, books. As a result, many were aware of the economic hardships these two scientists would have left behind at home. They might not have been paid for weeks—maybe even months.

The reminder had a dramatic effect. The air of embarrassed tolerance that had pervaded the room during the stumbling presentation quickly gave way to a wave of genuine sympathy. Thoughts of one-upmanship in the super-telescope league suddenly seemed uncomfortable—and rather unimportant. Thoughts of telescope closures also subsided as new possibilities for innovative and economical modes of operation began to suggest themselves. And, slowly at first, the preoccupation with aperture began to evaporate. The excitement of OWL remained undiminished, but its potential to rise above national boundaries seemed to take on new significance. Perspectives were restored; the fever abated. At last, 'Power Telescopes and Instrumentation into the New Millennium' had begun to find its soul.

Thus was the present and future of the telescope mapped out in the closing hours of Munich's epoch-making conference. A little while later, after all the goodbyes had been said, and promises to keep in touch and exchange information made, the last few delegates left the International Congress Centre. The late afternoon sunshine welcomed them back into the real world, inspired, no doubt, by what they had heard, and perhaps a little wiser. It felt pleasantly warm as they headed for home.

2

THE EYES OF DENMARK

PRELUDE TO THE TELESCOPE

It hardly bears thinking about, even today. Among all the dramas that have peppered the history of astronomy, few are more absurd than the after-dinner events at the home of a German professor of theology on 29 December 1566. Two hot-headed young men a long way from home, fanning a dispute that had been smouldering between them for days: tempers flared; swords were drawn. Their noble upbringing had trained them both well in the use of their weapons, and the impromptu duel they fought was swift and furious. Mercifully, it was also short. A diagonal slash to the face of one of the combatants, followed by the urgent entreaties of their dinner companions, ended the bout.

The bloody aftermath was a sorry young man with a deeply gashed forehead and the bridge of his nose missing. He was lucky to escape so lightly. The blade had come within half an inch of mortally fracturing his skull, and within an inch of taking his sight. Today, this man is remembered as the most capable scientist of his time, a man who used his abundant gifts—which included eagle-eyed visual acuity—to bring order to humanity's confused perception of the Universe. But how nearly had those gifts come to being snuffed out altogether on the wintry ground of a north German churchyard.

The duellists were Manderup Parsberg and Tyge Brahe. They were distantly related, both students, and both members of the Danish aristocracy. Thirty-five years later, in the gloomy aftermath of Tyge's untimely death, Manderup wished it to be placed on record that in spite of the injury he had inflicted on Tyge in the duel, the two had remained good friends throughout their lives. We have only his word for that, although the fact that one of his cousins had married Tyge's brother in 1580 lends it some support. It is also true that duelling was a common occurrence among the nobility in those violent days.

Of Manderup's life, we know quite a lot. Elevated in 1580 to the elite status of Councillor of the Realm, he played an important part in the government of Denmark. Of Tyge's life, though, we know vastly more. Here was no ordinary nobleman, not even by Manderup's high-ranking standard. Here instead was a man who spurned the traditional family pathway towards affairs of state in favour of the pursuit of knowledge. And so all-embracing were his talents that even now it is difficult for us to fathom the full extent of his contribution.

Today, we know Tyge by the Latin version of his name he adopted while a student at Copenhagen University in 1561—Tycho. Conventionally, it is pronounced 'Teeko' and his family name 'Bra-hay', but he would have referred to himself with a clipped and guttural 'Teuko-Bra'. He was born at Knutstorp, not far from the seaport town of Landskrona in what is now southern

Sweden, on 14 December 1546. He was the firstborn son of a large family.

Knutstorp is a leafy estate set in low, rolling hills some 20 kilometres inland from the strait that separates modern-day Sweden and Denmark—the Öresund. When Tycho was five years old, the ancient building in which he had been born was replaced by a fortified manor house of very generous proportions. It is still imposing today, even though the ravages of war (in the 1670s) and fire (in the 1950s) have diminished it considerably. It would have been a marvellous place to grow up in. But little, if any, of Tycho's childhood was spent there. In an event that seems unthinkable today, but could easily be rationalised in the unpredictable world of sixteenth-century Denmark, he was 'stolen' as a baby by his uncle Jørgen Brahe and aunt Inger Oxe, and raised by them as their son. There seems to have been little his natural parents could do but shrug their shoulders and get on with producing more children.

It was possibly his foster parents' influence, particularly Inger's, that steered Tycho away from a career in soldiering and service in the Danish Court towards more academic pursuits. That shift in emphasis first made itself felt when he completed his studies at Copenhagen University at the end of 1561. Instead of embarking on a tour of foreign courts, eventually to win his spurs as a knight, the fifteen-year-old Tycho began a series of visits to foreign universities. Though it could be justified as a means of acquiring an education in law, it was not the usual path of a courtier destined for government.

Tycho started with Leipzig, and it was here that he first found himself seriously drawn toward astronomy. Though he had studied the subject at Copenhagen with more than the usual enthusiasm (purchasing books far more advanced than his studies required, for example), it was only during his three years in Leipzig that the interest blossomed. And, as in the early lives of so many great historical figures, there was a distinctly clandestine element to it.

In those days before the telescope, astronomy concerned itself with the positions of celestial objects: the Sun and Moon, stars and planets. Where they appeared in the sky and how they moved over time were the only clues to their nature. True, the Sun was understood as the principal source of heat and light, and the Moon was thought to be a world not unlike our own—even if there was still a popular belief that it was shiny enough to reflect an image of the Earth's oceans and continents back to us. But what were the stars? And the planets that wandered through the stars—five of them, known since antiquity—what did their drunken peregrinations mean?

Today, almost all of us picture the Solar System in three dimensions: we imagine the Sun with all the planets—including our own—orbiting around it. Some of us allow our imagination to carry us beyond that, to the Sun as an inconsequential member of the vast catherine-wheel of stars and gas and dust that we call our Galaxy. But in Tycho's time, the sky was simply a surface, a stage on which the nightly performance was enacted. As far as most people were concerned, it might as well have been a huge hollow ball with the Earth at the middle.

Throughout history, a few individuals had looked for more than that, and had tried to conjure up a three-dimensional picture based on what could be observed. Ptolemy's view that the Sun, Moon and planets rotated about the Earth in a complex dance of superimposed circles had prevailed for 1400 years, but by the second half of the sixteenth century, Copernicus' new model of planets moving in circular orbits around the Sun was beginning to gain credence. It reflected

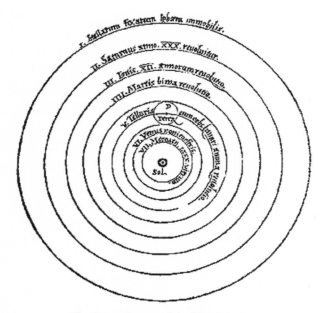

The Copernican model of the universe.

the ongoing spirit of the Renaissance, in which new thinking was replacing ancient dogma. It was by no means universally accepted, however, and in some parts of Europe—including Copernicus' native Poland—the Earth-centred theory of the Solar System was taught well into the eighteenth century.

What was needed in order to sort out the true nature of the Solar System was an accurate way of measuring the positions of the planets in the sky as they moved among the 'fixed' stars. By making such measurements over a long period of time, it would be possible to test the various models by seeing how well the planets' behaviour conformed to them. That quest eventually became Tycho's life's work—although it was only after his death that the answer finally came.

The secret pursuits that the youthful Tycho embarked on in

Leipzig in 1562 were already focused on this issue. He bought and read books on astronomy. He bought a small, easily-concealed celestial globe showing all the constellations of the sky. He bought tables predicting the positions of the planets from both the Ptolemaic and Copernican theories. And, even though his observational methods were the crudest imaginable—involving a piece of string held at arm's length to measure the angular distances between heavenly bodies—he was able to deduce that both these theories were in error. He was already on his way.

There was one downside to Tycho's fascination with the sky, and it is possible it was the same issue that ignited the fateful confrontation with Manderup Parsberg four years later. It was astrology. In the sixteenth century, astrology ranked alongside astronomy as a legitimate scientific pursuit, and any astronomer worth his salt was also expected to be able to cast horoscopes for the rich and famous—or anyone else, for that matter. This Tycho did as a matter of course, and as his involvement with astronomy grew, so also did the expectations of others as to his astrological prowess.

By the middle of 1566, Tycho had completed his studies in Leipzig, spent a year back in Denmark, and embarked on a new foreign tour. This time, he had gone to Wittenberg on the River Elbe. His stay was cut short after only five months, though, by the outbreak of plague, and by the end of September he had arrived in the north German university town of Rostock. Here, he matriculated at the university, and settled back into his studies of astronomy and astrology. And it was not long before astrology brought him trouble.

An eclipse of the Moon on 28 October was regarded as a significant event, and the nineteen-year-old Tycho's considered opinion was that it foretold the death of Suleiman the Magnificent, Sultan of the Ottoman Empire. Since Suleiman was then aged 70, this would probably have seemed like a fairly safe bet. Tycho publicly announced his conclusion, but was quickly embarrassed to discover that Suleiman's death had actually taken place several weeks before the eclipse. Ridicule followed—and no doubt it followed Tycho around for some time afterwards. Was it still in the air when Tycho and Manderup met at an engagement party on 10 December? Were they still disgruntled with one another when they met again at Christmas, and yet again on the fateful night? We shall never know, but it does seem a likely scenario.

The disfiguring wounds that Tycho received that night took time to heal. Perhaps it was during this period that his interest in medicine flowered. And perhaps, too, the inventiveness that led him to develop his own prosthetic noses (copper for weekdays and an alloy of silver and gold for Sundays, we are told) stimulated his interest in chemistry. Whatever their origins, those interests remained with him for the rest of his life. Tycho the astronomer was becoming Tycho the polymath.

Tycho Brahe, Denmark's lord of the stars.

HVEN

On a sunny day, the island of Hven sits like a green and brown jewel in the blue waters of the Öresund. Technically, its name has been Ven since 1959, when it was brought under the municipality of Landskrona, but the old Danish spelling is still commonly used. Either way, its name sounds something like 'vain' to the English speaker.

The island is small (4.5 kilometres long by 2.4 kilometres wide) and low, standing only 45 metres above sea level at its highest point. Its population of 362 is far outstripped by the 800 bicycles it offers for hire to visiting tourists. Yet there are six separate villages and a host of small farmhouses. In early summer, trim fields of green root crops and brilliant swathes of yellow rape remind one that this is primarily an agricultural landscape. It is not, however, rural charm that is the tourist drawcard.

In the very centre of the island, straddling the narrow road that joins the ports of Bäckviken in the east and Kyrkbacken in the west, lie two sets of partially restored ruins. On the northern side of the road are great earthen walls, clad in stone and forming a strictly symmetrical enclosure. Within its 78 metre square compass are gravel pathways, delightful gardens, a little Dutch Renaissance-style pergola, and an empty space in the middle. On the other side of the road is another plot with the same strict symmetry as the first, but only 18 metres to a side, and delineated this time by a wooden fence. The ruins within this boundary are not actually visible from the outside, but are protected by modern copper-clad enclosures, the largest of which is an imposing hemispherical dome.

These two ancient remnants are the sites of Tycho's house and his observatory. The larger is Uraniborg, the 'Castle of Urania', named in honour of the muse of astronomy. It served not just as Tycho's home, but also as the home of his household and the scientific community he had gathered around him. It contained his study, his library, his chemical laboratories and many of his observatory instruments. It was, to all intents and purposes, a scientific institution—and one as significant in its day as a Royal Observatory or a National Laboratory.

The smaller plot was a later addition. But, if anything, it was closer to the noble lord's heart. This was Stjerneborg, Tycho's 'Castle of the Stars'. The instruments it contained were the most accurate that had ever been devised for measuring the positions of heavenly bodies, their foundations sunk deep into the ground to gain a level of stability that Uraniborg's elevated floors could not match. Truly, it was the Cerro Paranal of its time. Nothing like it had ever been seen before.

Today, all that remains of Uraniborg are the earthworks enclosing the grounds. Where the castle itself stood is an empty space, startling in its smallness. Parts of the gardens have been lovingly restored and, at the easternmost corner of the enclosure, the man himself gazes eternally towards the

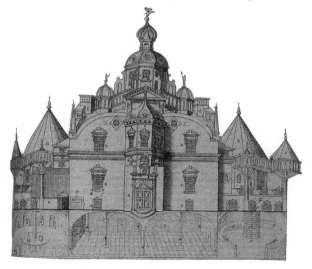

Uraniborg, Tycho's castle of Urania, as pictured in his *Astronomiae instauratae mechanica*, 1598.

skies of Hven in a twice-lifesize statue. His observatory, being partially underground, has fared better than his castle. Beneath the copper-clad enclosures remain the stepped circular depressions that allowed a standing observer to place his eye against a sight, no matter where in the sky his instrument was pointing. The brick paving on which the master himself walked is here; it is a far more evocative memorial than the modern statue.

Tycho celebrated Uraniborg's foundation at a cornerstone-laying ceremony on 8 August 1576, nearly a decade after the duel. His career had moved on rapidly since his student days. A sojourn in Denmark after his fateful stay in Rostock had seen the first glimmer of reconciliation of Tycho's family to his interest in science—although the process remained difficult and protracted. In 1568, a third foreign tour had taken him to Augsburg, centre of excellence in instrument-building. Here, Tycho had produced his first sighting instruments for accurately measuring the positions of heavenly bodies, developing his skills alongside the craftsmen of the city. Then, four and a half months after his return to Denmark at the end of 1570, his natural father died, leaving Tycho a wealthy Lord and master of his own destiny.

But destiny itself had also taken a hand for, on 11 November 1572, a new star had appeared in the northern constellation of Cassiopaeia. Bright enough to be seen in daylight, it was what we know today as a supernova—a distant and very massive star exploding with cataclysmic violence at the end of its life. Supernovae bright enough to be visible to the naked eye are rare events. Although another occurred in 1604, there was a long interval before the next one—the famous southern hemisphere supernova of 1987.

Over a period of months, the new star of 1572 faded back to insignificance, but not before Tycho had been able to determine that its position was fixed in the heavens—unlike the planets. This observation ran counter to the ancient wisdom that the sphere of stars is perfect and unchangeable. When, in 1573, Tycho published his discovery in a little book whose lengthy Latin title we now abbreviate to *De stella nova*, it placed the young scientist firmly on the road to international fame.

There had been one more false start before the edifice of Uraniborg began to take shape. Tycho's growing responsibilities to the Danish court, together with time-consuming duties at Copenhagen University and elsewhere, had led him to wonder whether his scientific career might not be better pursued away from Denmark. A fourth tour of Continental Europe in 1575 convinced him, and he began to make plans to settle in Basel, on the upper Rhine.

An audience with his monarch, however, resulted in an offer he couldn't refuse. Frederick II, 42-year-old king of Denmark and Norway, was preoccupied with his new castle at Helsinger on the western shore of the Öresund (the Elsinore Castle of Shakespeare's *Hamlet*), but was astute enough to realise that Denmark could ill afford to lose the services of this gifted young man. From the castle, Frederick could see the little island of Hven, and thought it would make a perfect spot for Tycho to pursue his scientific investigations. He offered it to Tycho, together with generous, lifelong royal patronage. Tycho accepted. It was an ideal solution.

With Uraniborg's completion in 1580, and more especially with the completion of Stjerneborg in 1586, Tycho entered the most productive period of his life. He built a succession of large astronomical instruments and, with them, he and his assistants observed all the phenomena of the sky, from transient events such as comets and eclipses to the motions of the Sun, Moon and planets—and of course, the positions of the stars.

The instruments all made use of the unaided eye. While their shapes varied according to the particular purpose for which Tycho had designed them, they all contained three basic ingredients: sights with which to align them on celestial objects, pivots or axes to enable them to be directed around the sky, and accurately engraved scales to allow the positions to be read off. Like the entire history of the telescope in microcosm, Tycho's instruments demonstrated a steady improvement in accuracy from one new model to the next.

A critical limitation of the instruments was the resolution of the human eye, the finest detail it can perceive. This is partly determined by the 'aperture' of the eye—the pupil diameter—but that is not a fixed quantity. The pupil grows as it adapts to the dark, reaching a maximum diameter of 7 mm (though there is a progressive inability to reach that size beyond the age of about 40). In the terminology used in Chapter 1, the intrinsic resolution of a 7 mm aperture is approximately 20 arcseconds.

In fact, the eye's resolution under most circumstances is determined by the size of the individual light receptors on the retina and is rather poorer than that—about 60 arcseconds, or 1 arcminute. Nevertheless, with cleverly designed instruments, repeated observations and corrections for such subtle effects as the bending of starlight in the air (atmospheric refraction), Tycho managed to achieve unprecedented positional accuracy. Frequently, it was as good as 25 arcseconds—at least ten times better than anything that had been achieved before.

The largest and most spectacular of Tycho's instruments on Hven—and one of the most accurate—was erected at Stjerneborg in 1585. Called the Great Equatorial Armillary, it was built with one of its axes aligned parallel to the axis of the Earth. This meant that an observer had to turn it only about that axis to maintain a star within his sights as the Earth rotated, a clever idea that was eventually applied to telescopes. In the case of Tycho's armillary, however, it would not have been a trivial task; its moving circle of wood, brass and steel was almost 3 metres in diameter, and very heavy.

A hint of this instrument's grandeur still remains at Stjerneborg. The reconstructed hemispherical dome on the site is about the same size as the original enclosure. It looks for all the world like a modern observatory building. But all that remains underneath is the empty, stepped crypt in which the great armillary was mounted.

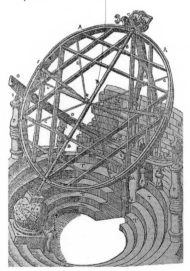

Tycho's Great Equatorial Armillary of 1585, the largest instrument at Stjerneborg.

LEGACY

Notwithstanding the noble Lord's wide-ranging scientific interests, he concentrated ever more diligently on the job of mapping the skies and charting the motions of the planets, Sun and Moon. Through the 1580s and early 1590s, ably supported by his 'family' of assistants and collaborators, Tycho Brahe built up a coherent set of high-precision observations that was the glittering jewel of sixteenth-century astronomy. This was his legacy to science, and its impact—its posthumous impact, as it turned out—was Earth-shattering.

Back in 1509, the 57-year-old Leonardo da Vinci—perhaps the greatest Renaissance polymath—had been utterly confident that our planet was at the very centre of things. We can judge his certainty from a diagram of the Earth, Sun and Moon in the *Codex Leicester*, a notebook dating from about that year now owned by software billionaire Bill Gates and his wife Melinda. In his diagram, Leonardo represents the orbits of the heavenly bodies as circles, and the conviction with which he has centred them on the Earth is almost startling. It is revealed by the bold hole made by the point of his compass, preserved for all time in the original manuscript.

But by 1609, with the publication of a book called *Astronomia nova*, the academic world had been told with compelling authority that the planets move, not in circular paths around the Earth, but in elliptical paths around the Sun. Our modern picture of the Solar System had been put in place, and the person who put it there—the author of *Astronomia nova*—was a brilliant, if rather ill-adjusted mathematician called Johannes Kepler. It is to Tycho Brahe that Kepler owes his place in history, for Tycho's observations—particularly of the planet Mars—were the raw material that allowed Kepler to formulate his laws of planetary motion.

Johannes Kepler was born on 27 December 1571 in Swabia, in southern Germany. His genius manifested itself at an early age, and by February 1600 he was working as a collaborator alongside the great nobleman himself. He was attracted to Copernicus' Sun-centred theory of the Solar System, rejecting a curious hybrid version devised by Tycho in which the planets revolve about the Sun—which in turn revolves about the Earth. Tycho had, however, made the radical deduction that comets can move in oval orbits, and this paved the way for Kepler's visionary interpretation of the Uraniborg observations. Using the methodology and data laid down by the astronomer of Hven, Kepler solved the problem of the Solar System once and for all.

Of Johannes Kepler's contribution to the telescope we shall hear more, but what of Tycho's? What exactly is it that qualifies this man to assume such a prominent position in the story of the telescope when he himself seems to have known nothing of it? His contribution was not really a technological one, except perhaps in the design of mountings for large astronomical instruments—the concept of an equatorial mounting that allows celestial objects to be followed around the sky by moving a single axis, for example. We might even question the extent to which Tycho would have made use of a telescope had he had one. In the 1670s, well into the era of telescopic observation, the great Polish astronomer Johannes Hevelius spurned the use of telescopes when making accurate positional measurements of the sky. Possibly Tycho would have done the same, although, like Hevelius, he surely would have exploited the telescope's power to reveal hidden detail, perhaps even rivalling Galileo in his achievements.

What sets Tycho Brahe apart from his predecessors in the context of the telescope is something far less tangible than technical innovation. It is to do with his global view of the problem in hand, and the way he mustered his resources to tackle it. He was the first modern-style director of a scientific institution, a man who could combine a deep professional understanding of his subject with consummate organisational skills to set up a major international facility—and then staff it with co-workers who were themselves leaders in their field. They, in turn, ensured that Tycho's influence would continue to reverberate throughout astronomy for decades after his death.

Tycho was not a man to let obstacles stand in his way. Realising that the successful promulgation of scientific discoveries hinges on the dissemination of the written word, he founded a printing press at Uraniborg. When shortages of paper threatened to disrupt the production of books and other publications, he built a large paper mill, sculpting the southern half of Hven into a system of dams and reservoirs to provide water for the mill's power source—a 7 metre diameter overshot waterwheel.

In his successful project management of major scientific infrastructure, Tycho set the standard for the great achievers of the telescope-building era to come—the Herschels, Rosses and Hales of this world. And perhaps it is not too far-fetched to suggest that if by some quirk of physics a freakish wormhole in space–time could have whisked Tycho from sixteenth-century Hven to turn-of-the-millennium Munich, he would have felt quite at home. In the heady atmosphere of 'Power Telescopes and Instrumentation into the New Millennium', he would have recognised instantly the ethos of large-scale scientific planning that he himself had pioneered. The only difference would have been that, to Tycho, the telescope was a human eye.

On 4 April 1588, Frederick II, king of Denmark and Norway, passed away. He was succeeded by his ten-year-old son, Christian IV. Imperceptibly at first, but with gathering momentum, Tycho Brahe's world on Hven began to fall apart. In a measure, it was due to his own lack of fastidiousness in maintaining his formal responsibilities to the Crown, but hostile courtiers close to the young King played perhaps the bigger part. Funding for Tycho's endeavours on Hven began to dry up, and eventually—inevitably—his annual pension was withdrawn. For all his scholarliness and eminence, Tycho still had a short fuse and, at the end of March 1597, he moved his entire household to Copenhagen, threatening to leave Denmark altogether if his royal patronage was not restored.

If the threat was a bluff, it didn't work. Accusations of oppression against the villagers of Hven followed the withdrawal of favours, and then came a court case probing violations of ordinances in the island's parish church. Tycho's world was crumbling around him. Early in June 1597, he left Denmark for Germany. He never returned.

Rostock, scene of the duel thirty years earlier, was Tycho's initial destination. He sojourned there for three months with his family, taking stock of his situation and writing a long letter to his monarch that did nothing to improve the situation. A move to Wandsburg Castle near Hamburg followed, where Tycho set up home as the guest of his scholarly friend Heinrich Rantzau, Viceroy of Schleswig-Holstein. Here some semblance of order returned to Tycho's life, but he was still without the patronage crucial to the continuation of his work.

It was only in 1599, after much careful diplomacy, that royal patronage was restored. It came from perhaps the most powerful ruler in Europe, the Holy Roman Emperor Rudolph II, and it eventually took Tycho to Prague. Such was Rudolph's enthusiasm for the great astronomer's work that it brought the promise of a brilliant new chapter in his life. But it was to be short-lived.

After a dinner party on 13 October 1601, Tycho found himself unable to pass water. Probably, it was due to hypertrophy of the prostate, although its onset was unusually sudden. Today, the problem would be fixed quickly and easily with a catheter, but for Tycho it spelled days of excruciating pain and increasing sickness as uraemia took hold. Eventually, on 24 October, he died. Modern studies have found high levels of mercury and lead in samples of Tycho's hair and beard, removed when his body was exhumed for autopsy in 1901. It seems likely that they are the result of his own medicaments, and that they contributed to his death. In the end, the physician could not heal himself.

Tycho was buried with great ceremony in Prague's Teyn Cathedral on 4 November 1601. A saying popular when Tycho had left his native land—'Denmark has lost her eyes'—now applied to the whole of Europe, and the world of learning felt the loss grievously. As for Tycho's family, they were devastated. There is no question but that Tycho had been a loving husband and father; his marriage to Kirsten Jørgensdatter—though problematic because of her commoner status—was one of the great success stories of his life. His children adored him. To them, Tycho had seemed the embodiment of all that is noble and good.

But there remains a nagging doubt about one issue that we would consider more important today than Tycho evidently did. That is his treatment of the villagers of Hven. It seems very likely that Tycho was a hard taskmaster, though it is difficult for us to judge the extent of any wrongful dealings. What is certain, though, is the speed with which the islanders took matters into their own hands once Tycho's enterprise there had been wound up by his bailiff. The bricks and stones of the castle, the observatory, the printing works and the paper mill were removed, recycled for use in the humble dwellings of the people.

Reports issued little more than half a century after Tycho's death speak of the complete disappearance of the buildings. Uraniborg and Stjerneborg were no more. On the island of Hven, the epoch-making works of Tycho Brahe faded as quickly as the morning mist lifts from the waters of the Öresund.

3

ENIGMA

WHISPERS OF ANCIENT TELESCOPES

Thanks to the copious records and notes that Tycho left behind, we have a very detailed picture of most aspects of his life. Certainly, his astronomical work is documented with painstaking thoroughness. We also know that he was a prolific communicator. Apart from his interactions with the colleagues gathered around him at Uraniborg and later in Prague, he corresponded frequently and widely with scholars throughout Europe. It was this aspect of his character that had led to his early recognition as a scientist, and he remained keenly aware of new developments.

History acknowledges Tycho as the leading authority of his day on the design and manufacture of astronomical instruments. Truly, the ancestral line of today's great telescopes goes back right through the observing facilities of Hven—from the Great Equatorial Armillary downwards. But nowhere in Tycho's writings is there any hint of an instrument that could possibly be described as a telescope. It simply doesn't figure.

The inference is, therefore, that it was unknown until it burst on to the stage of recorded history in 1608, seven years after Tycho's death. Yet, persistently—insistently—there are whispers that echo down to us from Tycho's time and before that speak of the existence of the telescope. What, for example, should we make of the following?

> But to leave these celestiall causes and things doone of antiquitie long ago, my father by his continual painfull practises, assisted with demonstrations Mathematicall, was able, and sundrie times hath by proportionall Glasses duely situate in convenient angles, not onely discovered things farre off, read letters, numbered peeces of money with the very coyne and superscription thereof, cast by some of his freends of purpose uppon Downes in open fieldes, but also seven myles of[f] declared wat hath beene doon at that instante in private places . . .

These words were written in about 1570 by an Englishman, Thomas Digges, who was an exact contemporary of Tycho—and also one of his correspondents. He died younger, in 1595. Digges wrote the passage in the preface of a book—begun by his late father, Leonard (c.1520–c.1559), and completed by himself—whose long, rambling title is usually abbreviated to *Pantometria*. It was concerned with the application of mathematics to such activities as surveying, navigation and gunnery. This passage and others in the book clearly refer to the idea of a telescope—an instrument that will bring an enlarged view of a distant scene to the eye. And we are told that this is no mere theoretical exploit, but the result of his father's 'continual painfull practises'.

Like skinning the proverbial cat, there are many ways to make a telescope. The simplest method requires either a combination of two glass lenses—one close to the eye, the other towards the object—or a combination of a dish-like concave mirror and a lens, usually with another subsidiary mirror to fold up the light path. These two forms can be seen today in most ordinary camera shops. In their modern incarnations they are highly developed products of the consumer optics industry, but they retain the same basic ingredients. The lens type, technically known as a refracting telescope, most often appears in pairs as binoculars, while the mirror type, or reflecting telescope, usually reposes seductively on a movable stand to tempt would-be amateur astronomers.

From *Pantometria*, and from the writings of another Elizabethan mathematician, William Bourne (also a contemporary of Thomas Digges), it is clear that the instrument

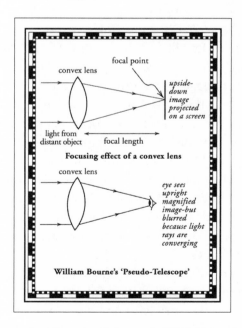

William Bourne's rudimentary 'telescope' of 1578. Unless the eye is long-sighted, a blurred image will result.

described by the two Diggeses incorporated a dished mirror—though probably in a manner quite unlike the modern reflecting telescope. Nevertheless, it has led some modern researchers—most notably the late Colin Ronan—to attribute the invention of the reflecting telescope to Leonard Digges sometime between 1540 and 1559, an idea that has provoked lively debate among historians. But what might be called mainstream scholarship upholds an opposing view, namely that the first practical reflecting telescope was constructed by Sir Isaac Newton in 1668—more than a century later.

There is a very good reason for this. The reflector's leisurely arrival into the world compared with the refracting telescope's dramatic appearance in 1608 is because it is much harder to make an adequate mirror than an adequate lens. (We shall see why in Chapter 7.) It is therefore most unlikely, particularly given some of the telltale details recorded by Bourne, that a successful instrument was built by Leonard Digges. He probably built something with features in common with a reflecting telescope, but the optical performance needed to reveal the denomination of coins cast about in open fields, or private goings-on 'seven myles' (11 kilometres) away, is far beyond the technology of the time. And, for that matter, of Newton's time.

Thomas' description in *Pantometria* is fantasy—wishful thinking, perhaps. And, for all that Bourne shines through his own words as an intelligent, likeable chap, his awed account of the Diggeses' achievements betrays him as a credulous and unreliable witness.

This is a pity, as elsewhere in William Bourne's writings there is a more sober account of the behaviour of lenses that hints at a rudimentary telescope. The reason it is not greeted with acclaim by historians is that, in order to use it effectively, an observer would have to have had defective vision.

The one glasse that must be made of purpose [says Bourne] is like the small burning glasses of that kind of glasse, and must bee round, and set in a frame as those bee, but that it must bee made very large, of a foote, or 14. or 16. inches broade, and the broader the better: and the propertie of this glasse, is this, if that you doo behold any thing thorow the glasse, then your eye being neare unto it, it sheweth it selfe according unto the thing, but as you doo goe backwardes, the thing sheweth bigger and bigger, untill that the thing shall seeme of a monstrous bignesse . . .

Remarkably, this is an exact description of something that was marketed 350 years later by a reputable London optical company as a 'window telescope'. Hanging in the window of the fashionable 1920s drawing room, this 15 inch (38 cm) diameter lens allowed 'a person standing a few feet within the room [to]

A window telescope from the 1920s.

see through the lens the details of the distant landscape . . .
From one's hall one may recognise the visitor far down the
drive. Or one may watch birds and other creatures . . .'.

The way this so-called telescope worked was crude in the
extreme. School physics reminds us that a convex lens—one
that is thicker in the middle than at the edges—will project an
upside-down image of a distant scene on to a flat screen like a
piece of card. It does that by changing parallel rays of light
from far-off objects into converging rays, and the position at
which those rays converge to a point is where the image is in
distinct focus. The card is then said to be at the focal point,
and it is separated from the lens by the focal length.

If, instead of a card, the eye is placed at the focal point, the
view through the lens is a confused blur—'like a myst, or
water', as Bourne puts it. But if the eye is moved towards the
lens, a magnified image of the distant scene appears. It is still
blurred because eyes aren't meant to accept converging light
rays—they much prefer parallel or diverging rays. However, if
the degree of convergence is slight (as it will be if the lens has a
long focal length) and *especially* if the observer is long-sighted
(which means that converging beams can be tolerated), an
acceptable, magnified, upright image of the distant scene will
be visible. The magnification is hardly 'monstrous'—objects
will appear no more than two or three times the size they are
with the unaided eye—but it is significant. The lens has
become a telescope—of sorts.

Bourne's account led the great mid-twentieth-century historian of the telescope, Henry King, to conclude that he—
Bourne—was long-sighted (hypermetropic), and this may
indeed have been the case. It is the usual fate of ageing eyes to
become hypermetropic. Unfortunately, Bourne's own description of his sixteenth-century window telescope then goes on to

introduce 'a very fayre large looking glasse that is well polished' in a manner that is completely confusing, begging the question as to whether he really was reporting accurately things he had done and observed. Like the answers to so many other questions surrounding the prehistory of the telescope, we shall probably never know.

CLOTHING THE NAKED EYE

The writings of another sixteenth-century optician have been famously interpreted as including a description of a telescope. But here we move towards optical components on a scale quite different from the lens that Bourne described. While large burning-glasses (crudely fashioned convex lenses of short focal length) might have been quite familiar in the sixteenth century, Bourne's large, long focal-length lens would have been a real rarity. Its optical quality—the accuracy of its surfaces and the homogeneity of the glass—must have been very poor. Of somewhat better quality, because they were smaller and more widely manufactured, were spectacle lenses.

Like the origin of the telescope, the early history of spectacles is surrounded by mystery and speculation. The best scholarship indicates that spectacles made their European début in Italy towards the end of the thirteenth century as a remedy for presbyopia, the inability to focus on nearby objects that accompanies advancing years and usually culminates in long-sightedness like Bourne's. Convex lenses are needed to compensate for this; they simply act as magnifying glasses, replacing the lost focusing power of the eyes. Take two matched convex lenses, mount them side by side in a frame and—*voilà!*—you have your spectacles. If you are an ageing monk painstakingly transcribing illuminated manuscripts, you will hail them as a godsend.

Concave lenses—those that are thinner in the middle than at the edge—are required for a different visual complaint, one that afflicts the young as well as the old. Sufferers from myopia, or short-sightedness, can only focus on distant objects if concave lenses are used to transform parallel rays of light into divergent rays. In spectacles, they made their entrance rather later than convex lenses because they are harder to make. Again, it was in Italy, but probably around 1450. Their appearance in northern Europe was later still.

It is the work of an Italian optician, Giovanbaptista Della Porta (1538–1615), that quickens the pulse of anyone wishing to place the invention of the telescope before the seventeenth century. This man wrote a monumental bestseller named *Magia naturalis* (quite simply, *Natural Magic*), published in 1589, which includes the following passage:

> Concave Lenticulars [lenses] will make one see most clearly things that are afar off; but Convexes, things neer hand; so you may use them as your sight requires. With a Concave you shall see small things afar off, very clearly; with a Convex, things neerer to be greater, but more obscurely: if you know how to fit them both together, you shall see both things afar off, and things neer hand, both greater and clearly. I have much helped some of my friends, who saw things afar off, weakly; and what was neer, confusedly, that they might see all things clearly.

The reason this passage arouses so much excitement is that when the telescope finally made its undisputed appearance, towards the end of 1608, it was in a form that took a convex and a concave lens and 'fit them both together'. The eventual layout was similar to Bourne's 'window telescope' (although on a much smaller scale), with the eye placed just inside the focal point of a convex lens. But the additional component that turned it into a genuine telescope was a steeply curved concave lens immediately in front of the eye. The effect of this lens is to render a normal eye artificially long-sighted—which is exactly what is needed to see a sharp, upright, magnified image in the convex lens. Today, we call this combination of lenses a Galilean telescope, even though Galileo's role was improver rather than inventor. It still survives in the guise of ordinary opera glasses.

It is the context of Della Porta's comment that pours cold water on the idea that he is describing a telescope. He is clearly discussing the correction of defective vision rather than a version of Digges' 'proportionall Glasses'. This point has been made before by Albert Van Helden, one of the most fastidious modern researchers on the early telescope. Van Helden interprets Della Porta's description as being of a weak Galilean telescope to help people with extremely poor eyesight. And it is true that ungainly spectacles using this principle eventually became fairly common, until more sophisticated therapies emerged during the second half of the twentieth century.

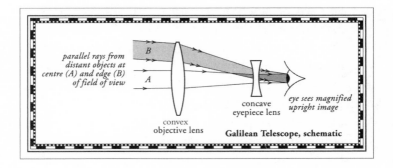

parallel rays from distant objects at centre (A) and edge (B) of field of view

B

A

convex objective lens

concave eyepiece lens

eye sees magnified upright image

Galilean Telescope, schematic

In the first true telescope, the small concave lens renders a normal eye artificially long-sighted, perfecting the image.

However, there is an alternative interpretation. This is that Della Porta was actually describing rudimentary bifocal lenses. Cut in half and 'fit together', they would provide an unusual aid for mildly myopic individuals needing help with both near and far sight. Bifocal lenses are usually assumed to have been invented by Benjamin Franklin during the 1750s, but the possibility that Della Porta could have experimented with them before 1589 does seem to be suggested in the passage from *Magia naturalis*.

Either way, no telescope in the usual sense of the word is implied. And this conclusion is not really surprising, given the military and scientific potential of a workable telescope. Had anyone truly managed to make one, the news would have spread like wildfire—as it did in 1608—and it would have been rapidly exploited—as it was in 1609. The fact that such events are not recorded in the sixteenth century leads us to the inevitable conclusion that stories of amazing instruments, like the Diggeses' account, are just stories.

Perhaps the most sober comments on the sixteenth-century telescope come from one of the greatest of all Renaissance figures, Leonardo da Vinci (1452–1519). Leonardo's legendary artistic ability was at least partly due to his keen-eyed observations of the natural world. His soaring achievements in science and technology owed much to them, too. He was naturally interested in the behaviour of light, and thought deeply about such questions as whether the atmosphere bends starlight, and why the sky is blue. He didn't always arrive at the right answers, but he got remarkably close given the scientific tools at his disposal. In one of his astronomical observations, at least, he was spot-on: he was the first person to deduce that the faintly illuminated lunar disc seen in the crescent Moon is due to light reflected from the Earth.

The unmistakeable idea of a telescope figures several times in Leonardo's notebooks, but it is always couched in hypothetical terms. And while he may have come close to discovering the recipe for the Galilean instrument, in the end he evidently didn't. Had he made one, he of all people would have exploited the telescope to the full, recording his discoveries in his famous mirror-writing. But we also have his own memorandum to himself—preserved in the *Codex Atlanticus* in Milan—that betrays his lack of success. 'Fa ochiale da vedere la luna grande', it reads ('Construct glasses to see the Moon magnified').

In some respects, the natural philosophers of the Renaissance and the later sixteenth century had a vested interest in claiming to have invented the telescope. The device was the Holy Grail of optics, there being a common assumption that it had been known in ancient times, but that the knowledge had been lost. Indeed, one of the eventual petitioners for a patent on the telescope in October 1608, Jacob Metius, said in his patent application that he had been investigating 'some hidden knowledge which may have been attained by certain ancients through the use of glass'.

The search for such hidden knowledge is as fascinating today as it was then, but if the evidence of sixteenth-century writers is so difficult for us to interpret, how much more so are the words of earlier authors—especially when any public declaration of interest in optical science was likely to result in accusations of necromancy, or black magic.

LEGENDS AND LENSES

Three hundred years before Della Porta wrote *Magia naturalis*, another scientist who has been credited by some with knowledge of the telescope languished in prison on exactly this charge. Roger Bacon (*c.*1214–1294), Franciscan friar and Oxford scholar, wrote his own 'great work', *Opus maius*, sometime around 1267 as a result of a bungled approach to Pope Clement IV. He had written to the pontiff in 1266 suggesting that the Church should coordinate the production of a great encyclopaedia of science. Unfortunately, Clement mistook the proposal for an announcement that Bacon had actually completed such a work, and enthusiastically asked to see it. Bacon was aghast. Realising that he could not disobey the pope, he rapidly produced three remarkable volumes, of which *Opus maius* was the first. One hopes Clement was suitably impressed.

Unfortunately for Roger Bacon, popes don't last for ever, and within a decade he was condemned by a Church under new management and imprisoned until shortly before his death. *Opus maius* contains much that was deemed heretical, but of interest to us is Bacon's suggestion that one might

> . . . so shape transparent bodies [lenses], and arrange them in such a way with respect to our sight and objects of vision, that the rays will be refracted and bent in any direction that we may desire, and under any angle we wish we shall see the object near or at a distance. Thus from an incredible distance we might read the smallest letters and number grains of dust and sand . . .

No doubt the sheer audacity of this passage enraged Church authorities. But in fact, Bacon is here echoing the words of one of his teachers, Robert Grosseteste (*c.*1168–1253) in a book about rainbows called *De Iride*. Grosseteste had written a number of scientific treatises and had evidently managed to steer clear of the cardinals, becoming bishop of Lincoln in 1235. From our perspective, the passage reveals the ability of both men to conceive of a telescope in terms of what it might do, and the kind of apparatus that could bring it about. But that is all. Other than these words and the later testimony of a sixteenth-century English scientist called Robert Recorde that Bacon's optical studies did not involve sorcery, there is nothing to substantiate claims that the learned friar actually knew how to make a telescope.

It had already been the fate of another early thinker to be accused of doing deals with the Devil. This man was only saved from imprisonment—or worse—by the fact that he was actually the pope. Nevertheless, he was viewed with deep suspicion by his contemporaries. Gerbert of Aurillac, a Frenchman, lived during the second half of the tenth century, acceding to the Papacy as Sylvester II in the year 999. His enthusiasm for technology—such as it then was—and, in particular, his introduction of the abacus to Europe, has led some modern writers to dub him 'the first millennium's Bill Gates'. Certainly, he had the stamp of a polymath, his eclectic interests ranging from music to mathematics, and from anatomy to astronomy.

Gerbert built an astronomical clock that incorporated sighting devices described as 'tubes', and the suggestion has been made that they were telescopes. This is hard to believe, given that Gerbert's track record in optics was unremarkable. Much more likely is the possibility that they were simply empty tubes, for it had been held since at least the time of Aristotle (384–322 BC)—and probably long before—that vision was improved by looking through a tube.

This idea is not as silly as it sounds. Confining the field of view of the eye to a small area reduces glare and enhances concentration on what is seen. It is an experiment easily performed, and worth the effort. 'Tube vision' does not, as is often suggested, allow stars to be seen in daylight, but it would certainly allow the direction of a star to be estimated at night in a crude forerunner of the accomplishments of Tycho Brahe. Such advantages might explain why representations of individuals holding sighting tubes to their eyes are occasionally found on ancient artefacts and documents—even on some that date back to the earliest flowering of civilisation.

So what of the ancients themselves? Could they *really* have known the secret of the telescope, only for it to be lost by succeeding generations? Did they even know about lenses? And who were these ancients anyway?

Elsewhere in Roger Bacon's *Opus maius* is the assertion that Julius Caesar scanned the English coastline from Gaul prior to his invasion of 54 BC using an optical device. Thus, the Romans are one ancient civilisation rumoured to have had telescopes. Then, there are extraordinary feats of signalling over the 134 kilometres separating north Africa from the western tip of Sicily, said by a second-century BC Greek historian to have been accomplished by invading Carthaginians some two hundred years before his own time. Do they suggest the use of telescopes? And earlier still, around 750 BC, ancient Assyrians recorded on cuneiform tablets that they had lenses and tubes of gold. Could the fact that the Assyrians always depicted Saturn as a god surrounded by a ring of serpents be telling us that their court astronomers had used primitive telescopes to observe the sky?

Perhaps most provocative of all is an obscure passage in the writings of the first-century BC Greek author Diodorus, which refers to a mythical race of Hyperboreans far to the north. The Hyperboreans saw the Moon as 'but a little distance from the Earth, and to have upon it prominences, like those of the Earth, which are visible to the eye'. There is a tacit assumption among some scholars that the Hyperboreans were none other than the ancient Britons. Is it remotely possible that the builders of Avebury and Stonehenge could have known about telescopes in the second millennium BC?

Intriguing though these accounts of ancient achievement are, there is really nothing to substantiate them. And difficulties of one kind or another accompany them all. Caesar's knowledge of the British coast, for example, has been attributed by other early historians as being due to spies rather than surveillance. Further doubt is cast on Roger Bacon's account by the fact that it mentions Caesar using mirrors to make his remarkable observations of the cities and camps of England. As we have already noted, curved mirrors suitable for use in telescopes are even harder to make than lenses.

In the same vein, optical signalling over scores of kilometres of rolling Mediterranean would have challenged early twentieth-century technology, let alone its fourth-century BC counterpart, while a similar objection can be levelled at anyone wishing to place the discovery of Saturn's rings earlier than the seventeenth century. Telescopes had to evolve quite a long way before the rings' true nature was revealed in 1659. And myths of Hyperborean moon-watchers? Well . . . great stories, these myths.

But there remains one slender thread of evidence that tantalises scholars of the telescope's prehistory, and begs the question as to whether someone, somewhere, might *just* have succeeded in making a device to see distant objects. It is that—yes—the ancients did have lenses. They knew how to make them. And many of them repose today in the ancient history museums of the world.

The most famous one is a 42 mm by 34 mm oval convex

lens, 6.2 mm thick in the middle and 4.1 mm thick at its edge, known variously as the Layard lens (from its discoverer, Austen Henry Layard, whose archaeological dig unearthed it in 1849) or the Nimrud lens (from the ancient Assyrian capital in whose ruins it was found). It has been reliably dated as no later than the seventh century BC, and it is in the possession of the British Museum. Needless to say, its existence is cited as clinching proof by enthusiasts of ancient Assyrian telescopes.

The Nimrud lens is made of rock crystal, a naturally occurring colourless quartz that can be shaped and polished like glass. Many ancient lenses are made of crystal, but glass ones, too, have been found. In fact, glass was quite well known in ancient times. Its manufacture was probably discovered by chance, and fragments dating back to the third millennium BC have been unearthed in Mesopotamia. But glass lenses, being chemically less stable than their natural counterparts, tend to become opaque on archaeological timescales, part-icularly when exposed to the air. Sometimes, there is little to distinguish them from smooth round pebbles.

Lenses have been unearthed in many of the great centres of the ancient world: Ephesus, Cairo, Carthage, Troy, Mycenae and the Roman colonies in Europe. Some have been found in Britain. And a significant hoard of Viking rock crystal lenses, dating from around the eleventh century AD, has been found on the island of Gotland in the Baltic Sea, not 400 kilometres from Tycho Brahe's Hven.

With very few exceptions, these lenses are all of the same type. They are magnifying glasses, convex lenses with a focal length of a few centimetres perfect for use by ageing eyes. Perhaps they enabled their owners to apply the microscopic decoration that adorns some ancient artefacts. The Nimrud lens belongs in the same category. As far as their use in telescopes is concerned, they are the wrong kind of lenses. None has been found with the reasonably long focal length needed to be the main, image-forming lens of a telescope—the so-called objective lens. And this is not surprising, since the shallow curves required on such a lens would be far beyond the skills of any ancient craftsman (see 'Why there are no ancient telescope lenses' overleaf).

Nevertheless, the existence of such lenses, coupled with legends like those mentioned above, has led some modern writers to turn the whispers of ancient telescopes into a hysterical shout. How, they demand, can any sane person ignore all this compelling evidence? In the cold light of reason, though, the evidence looks vague and highly circumstantial, and flies in the face of all we know about the development of optical technology.

Not surprisingly, the academic establishment doesn't like being shouted at. It will remain unconvinced of the pre-seventeenth-century telescope until scientifically verifiable evidence emerges: an intact telescope, perhaps, or an unambiguous drawing with incontrovertible provenance. From what we know already, it seems impossible that such artefacts will ever come to light. But what a profound impact they would have on our understanding of early technology if they did.

WHY THERE ARE NO ANCIENT TELESCOPE LENSES

Lenses are made by grinding them against a hard tool using successively finer abrasives. Sand was used in making the earliest lenses. The final polish is applied with a softer material such as leather (or, in modern times, pitch) using a very fine abrasive like jeweller's rouge. Controlling the curvature of the surface is not easy, and it was only towards the end of the sixteenth century that optical craftsmen developed techniques for producing the shallow curves needed on telescope objective lenses. By contrast, the steeply curved surfaces found on ancient lenses are exactly what would be expected from crude experimentation.

There is another, more subtle consideration. When a convex lens is used simply as a magnifying glass or spectacle lens, the individual pencils of light passing through it in any given direction are only as wide as the eye-pupil that receives them—a few millimetres at most. Inaccuracies in the lens such as variations in curvature over its surface show up simply as distortions as the eye looks in different directions. But in a telescope objective lens, pencils of light filling the whole of the lens contribute to the image at any point. Surface inaccuracies therefore degrade the quality of the entire image.

Any ancient lens used as a telescope objective would degrade the image far beyond acceptable limits. In fact, we know from surviving examples of early seventeenth-century telescopes that objective lenses were only just satisfactory even then. They still had to have their outer, more irregular parts masked off to work properly.

4

ENLIGHTENMENT

THE TELESCOPE MAKES ITS DEBUT

When the telescope finally became a reality, it was not at the hand of a great thinker—a Gerbert of Aurillac, a Leonardo da Vinci or a Tycho Brahe. Rather, it came from the work-stained hands of small-time craftsmen hoping to make a fast buck in a time of national crisis. These men were spectacle-makers, practical opticians, who plied their trade in early seventeenth-century Holland.

Ultimately, it was religious dispute that brought the telescope out of the woodwork—and then went on to provide a dramatic backdrop to its debut on the stage of recorded history. The fourteenth and fifteenth centuries had seen a growing level of protest against the Church in Rome because of perceived corruption in high places. Early in the sixteenth century, the protestors—or Protestants—formed a new church in the religious revolution we call the Reformation. It divided northern Europe from the staunchly Catholic south.

The seventeen provinces of the Netherlands—a part of the Holy Roman Empire—espoused Protestantism, goading Philip II of Spain to threaten the dreaded Inquisition in order to force the new converts to recant. The Dutch were having none of that, though, and in 1568, went to war with Spain. The ensuing long, drawn-out conflict is now known as the Eighty Years' War and, indeed, its final resolution did not occur until the Europe-wide Peace of Westphalia in 1648.

The southern provinces of the Netherlands—which form much of present-day Belgium—succumbed to Spanish pressure when the war was in its tenth year, but the seven northern provinces held out, forming themselves into a fledgling republic centred on The Hague in the province of Holland. These United Provinces were governed by an assembly called the States General whose leader was known as the Stadtholder and, of all the men who held this highest office in the northern Netherlands, none was more successful than Prince Maurice of Nassau. (Not to be confused with its better-known namesake in the Bahamas, Nassau in present-day Germany was then a principality of the Holy Roman Empire.) Maurice ruled from 1585 until his death in 1625, and it was into his powerful hands that the first verifiable specimen of a telescope fell during the last week of September 1608.

Neither the timing of this event nor the choice of recipient was accidental. Maurice was not only Stadtholder, but also commander-in-chief of the armed forces of the United Provinces, and he had shown himself to be a brilliant tactician in the war against the Spanish. He was also an able diplomat and, during much of 1608, had been exercising those skills in a long and complex series of peace negotiations brokered by the French. Truly, The Hague had become the Camp David of seventeenth-century Europe, and it was filled with the diplomatic delegations of many nations.

That they were eventually successful is borne out by the Twelve Years' Truce of 1609, but in September 1608, things looked black indeed. On the last day of that month, the talks ended in near-deadlock with the departure of the Spanish delegation led by Ambrogio Spinola, commander-in-chief of the Spanish forces in the southern Netherlands. How extraordinary, then, that during that same week, Prince Maurice should be presented out of the blue with a military device of enormous strategic significance—the first working telescope.

The man who turned up carrying the instrument was its maker, a 'humble, very religious and God-fearing man' (according to an ambassador from Siam) called Hans Lipperhey. He came from Middelburg, the principal town of the province of Zeeland which lay to the south of Holland. Lipperhey was actually German by birth, but had built up a flourishing business in Middelburg as a spectacle-maker and somehow had stumbled on the secret of the Galilean telescope—the simple but clever combination of lenses described in Chapter 3.

Whether this was by accident, by his own ingenuity or because someone else had shown him, we shall probably never know. Much has been written on this topic, but given the ease with which the discovery can be made once the right lenses are to hand, the accident hypothesis seems attractive. Personal experience bears this out. Decades ago, at the tender age of eleven, I was surprised and delighted to find the same secret when playing with two lenses left over from my father's home-made photographic enlarger. It was an almost inevitable discovery.

In any event, the key ingredient was the availability of a rather weak (long focal-length) convex lens to act as the objective and a rather powerful (short focal-length) concave lens for the so-called eyepiece. As we saw in Chapter 3, it was only by the turn of the seventeenth century that spectacle-makers had developed the techniques required to produce the shallow surfaces of the objective. Likewise, the steep curves of the eyepiece would have presented serious challenges to earlier opticians because of their concavity.

In that sense, therefore, the telescope was very much a product of its time. All Hans Lipperhey had to do was to select a couple of suitable lenses, fit them into a tube and set off for The Hague. With national security in the balance, this must have seemed like the perfect time to make a financial killing— no matter how humble, religious and God-fearing he was.

Lipperhey had prepared himself well, and he carried with him a letter from the Councillors of Zeeland to their representative at the States General in The Hague. 'Den brenger van dese . . .', it began:

> The bearer of this, who claims to have a certain device, by means of which all things at a very great distance can be seen as if they were nearby, by looking through glasses which he claims to be a new invention, would like to communicate the same to His Excellency [Prince Maurice]. Your Honour will please recommend him to His Excellency, and, as the occasion arises, be helpful to him according to what you think of the device . . .
>
> Honoured, etc., the XXVth September, 1608.
>
> > Herewith,
> > Councillors.

Subsequent events revealed that what Lipperhey was really interested in was a patent on his claimed invention, or perhaps a government pension for supplying it. But first he had to demonstrate that it worked and so, sometime between 25 and

30 September, the spectacle-maker was ushered into the presence of the great Stadtholder to present his new device. In order to test it, Prince Maurice climbed the tower of his residence in the grounds of the Binnenhof—the imposing thirteenth-century building that was then the seat of the States General and remains its ceremonial home today. From there, we are told, he could clearly see through the instrument the clock of Delft and the windows of the church of Leiden, respectively one-and-a-half and three-and-a-half hours' journey away.

In fact, the line-of-sight distances to these cities from The Hague are 8.6 kilometres and 17.6 kilometres. One might be excused for thinking that the report sounds suspiciously like those extravagant claims made for the hypothetical telescopes of the previous century. And indeed, the lenses Lipperhey had at his disposal would probably have restricted the magnification of his telescope to about three times, meaning that the two landmarks would have appeared through the telescope as if they were 3 and 6 kilometres distant respectively—still a long way off. However, these are *very* substantial buildings. And no one said anything about actually reading the time on the clock . . .

The moment in which Prince Maurice first held the telescope in his hands was an event of high drama, a turning point in history against a backdrop of international crisis. For the first time, technology had been able to extend one of the human senses, and today we perceive that moment as pregnant with all the potential consequences—military, scientific and philosophical. Yet, in the event, the drama was short-lived. Almost immediately, it dissolved into farce. The telescope's long-awaited debut on the world stage turned out to be in the starring role in a comedy of errors.

CLAIM AND COUNTERCLAIM

Whether it was by dint of diplomatic etiquette, a misguided ploy, or simply by mistake, we have no way of knowing. But somehow, the unthinkable happened. Within days of Prince Maurice inspecting the telescope—or perhaps even at the same time—he had shown it to the Marquis Spinola. By the time the commander-in-chief of the enemy forces left The Hague on 30 September, he had not only held the new strategic deterrent in his hands, but had taken a look through it.

His reaction was one of amazement. 'From now on, I'll no longer be safe, for you'll see me from afar,' he said to Prince Frederick Henry, Maurice's half-brother. 'Don't worry,' the Prince reassured him, 'we'll forbid our men to shoot at you.' No doubt everyone laughed.

On his arrival in loyalist Brussels a few days later, Spinola recounted the details of the new invention to his master, the Archduke Albert, and also to a high-ranking papal nuncio called Guido Bentivoglio. This man was in almost constant contact with Rome, and his correspondence is one possible route by which Galileo could have heard of the telescope. Indeed, in April 1609 Bentivoglio sent a specimen of the new invention to Rome. Maurice's disclosure, farcical though it was, may have had enormous consequences for the rapid scientific exploitation of the telescope.

Meanwhile, back in The Hague, the members of the States General were eager to see for themselves the marvellous new device that Prince Maurice had received. He sent it to them, commenting that with its help, 'they would see the tricks of the enemy'. He seems to have neglected to add that the same enemy already knew of its existence. Then, on Thursday 2 October, Lipperhey himself was interviewed by the States

The entry in the States General's Minute Book recording Hans Lipperhey's patent application for his telescope, 2 October 1608.

General at the Binnenhof. He put his cards on the table, asking them to grant him a patent for 30 years during which time no one else would be permitted to make telescopes. Failing that, he would be happy with a yearly pension in return for which he would make telescopes solely for the State.

The neat, measured handwriting of the duty clerk records for posterity the States General's predictable reaction. Even in those days, it was the time-honoured way of deciding what to do next. They formed a committee. They also made the curious request that Lipperhey should try to improve his invention so that it could be used with both eyes. While this might seem to us to betray a failure to appreciate the signifi-cance of what they were being offered, we have to remember that this was the first real optical instrument. The idea of peering through an eyepiece with one eye—familiar enough today—would have seemed unnatural and bizarre in the early seventeenth century.

It is also worth bearing in mind that as yet, the telescope had no name. Over the next few days, as the States General's committee debated the fate of Lipperhey's patent application, it was referred to variously as 'the instrument for seeing far', 'the instrument invented by Johan [sic] Lipperhey' and 'the invention to stretch out sight'. It was not until three and a half years later, when an exclusive group of Italian and Greek intellectuals held a banquet to honour Galileo and his astronomical discoveries, that the new instrument was christened *telescopium*—the far-seeing instrument. Names with a similar literal meaning emerged throughout Europe: *Fernrohr* in Germany and *verrekijker* in the Netherlands, for example. But across the English Channel, common usage produced a spate of outlandish names: perspective glass, spyglass, perspective cylinder and—in homage to its nation of origin—Dutch trunk. As always, British eccentricity shone through.

The minutes of the States General show that while the matter of Lipperhey's patent application was dealt with briskly enough, there was no urgency to grant his request. On Saturday 4 October, it was recorded that he would be asked to deliver up to six binocular (two-eyed) versions fitted with rock crystal lenses within a year. The next day, he was paid a first instalment of 300 guilders towards the cost of these ambitious instruments, with a further 600 guilders to be paid on delivery. Lipperhey was admonished not to reveal the secret of his invention to anyone, and no doubt felt some optimism that either his patent or his pension would eventually be granted.

Less than two weeks later, however, his aspirations were dashed as the affair once again lurched into the realm of farce.

On Friday 17 October, another man turned up at The Hague carrying a telescope and a letter to the States General. This time it was an instrument-maker from Alkmaar in the northern part of Holland, a man named Jacob Adriaenszoon—but more commonly known as Jacob Metius. He was a much more imposing figure than the humble Lipperhey: his father, Adriaen, was a former burgomaster of Alkmaar, and his brother (another Adriaen) was a professor of mathematics and astronomy who had studied with none other than Tycho Brahe.

Metius testified in his letter that:

> . . . he, the petitioner, having busied himself for a period of about two years, during the time left over from his principal occupation, with the investigation of some hidden knowledge which may have been attained by certain ancients through the use of glass, came to the discovery that by means of a certain instrument which he, the petitioner, was using for another purpose or intention, the sight of him who was using the same could be stretched out in such a manner that with it things could be seen very clearly which otherwise, because of the distance and remoteness of the places, could not be seen other than entirely obscurely and without recognition and clarity.

In other words, he had accidentally discovered the telescope. In similarly long-winded style, Metius went on to say that he had heard of the invention made by the spectacle-maker of Middelburg, that his own prototype instrument had been tested against that one, and that he, too deserved a patent on the invention because of his own 'ingenuity, great labour and care (through God's blessings)'.

The States General wisely refrained from appointing another committee, but awarded Metius 100 guilders, requesting him to 'work further in order to bring his invention to

Johannes Stradanus' depiction of a spectacle seller, whose wares are
obviously popular.

greater perfection, at which time a decision will be made on
his patent in the proper manner'.

If there was any doubt left in the minds of the members of
the States General that the invention was already too widely
known to be patented, it must have been dispelled altogether
by a third letter—this time unaccompanied—which arrived
from Middelburg at about the same time as Metius turned up.
Written on 14 October, it was from the same Councillors of
Zeeland who had provided Lipperhey's original letter of rec-
ommendation. In Middelburg, it said, there were now others
who knew the art of seeing far things and places as if nearby.
In particular, there was a young man who had demonstrated a
similar instrument to Lipperhey's—and what would the
States General like them to do about him?

OPEN SECRET

The States General's answer is not recorded, but it was quite clear that these new devices were popping up all over the place. The later (1614) testimony of another well-known astronomer and former associate of Tycho, Simon Marius, was that he first heard of the telescope from his aristocratic patron (a man with the unenviable name of Johann Philip Fuchs von Bimbach), who had seen one offered for sale at the Autumn Fair in Frankfurt in September 1608. The would-be vendor was a Dutchman who claimed to be the inventor; Fuchs didn't buy it because one of the lenses was cracked.

It is not easy to explain how a telescope could appear in Frankfurt, 500 kilometres from The Hague, during the same month that had seen Lipperhey's visit to Prince Maurice. Some scholars have suggested that the Dutchman at Frankfurt and the young man mentioned by the Councillors of Middelburg were one and the same person, and that it had taken this individual from the end of September until the second week in October to make his way home from the Fair.

There is even a hint as to his identity in the shape of an account by a man who claimed to be his son—one Johannes Sachariassen. In 1634, Johannes boasted to a friend that his father, Sacharias Janssen, had invented the telescope. Certainly, he had been a resident of Middelburg and a spectacle-maker. Unlike the models of integrity represented by Lipperhey and Metius, however, Janssen was an extremely shady character who had been in trouble with the law for debt, assault and forgery—for which he was eventually threatened with the death penalty. He had wisely disappeared before it could be carried out.

For what it is worth, another strand of evidence from Janssen's son suggests that the small-time crook had copied his telescope as early as 1604 from one that belonged to an Italian. It was said to have carried the inscription 'anno 1[5]90'. The historian Albert Van Helden has presented a scenario in which this mysterious instrument was one of the weak telescopic aids to vision that Giovanbaptista Della Porta might have been describing in his *Magia naturalis* of 1589 (see Chapter 3). Middelburg is known to have hosted large numbers of Italian exiles, most of whom were deserters—mercenary soldiers tired of helping the Spanish in their attempt to subdue the United Provinces. Perhaps Janssen had indeed made a copy, and managed to improve it to the stage where it could provide useful magnification.

The modern consensus is that we will never know for certain who actually invented the telescope. An answer to the question of why it emerged so suddenly from obscurity—apparently in several places at once—is similarly elusive. But, as we have seen, the necessary ingredients were all there: optical technology that had just developed to the stage where lenses of the required quality could be produced, the inevitability of the right combination of lenses being discovered once they were available, and the pressure of international crisis to draw the invention out of the hands of its makers and into the waiting grasp of politicians.

There are other snippets of evidence. We know, for example, that a handful of telescopes dating from the first 30 or so years of the seventeenth century and still in existence today used good-quality Venetian glass (intended for mirrors) as the raw material for their lenses. Perhaps it was a shipment of this glass to Holland that resulted in the observed cluster of Dutch telescopes in 1608.

There seems little doubt that several spectacle-makers knew the secret of the telescope in the early 1600s. While

Metius' claim to originality appears to be genuine, his device may not have been the first. Alkmaar was relatively remote in northern Holland. Middelburg, on the other hand, had a major glass factory of its own, the only one in Zeeland. That small Dutch town seems perhaps the more likely place for the first telescope to have seen the light of day.

On 11 December 1608, Hans Lipperhey brought to the States General the first of the binocular telescopes with rock crystal lenses that he had been commissioned to make. It was inspected by the members of the committee, and found to be good. Four days later, a further payment of 300 guilders was made to him, and he was asked to produce two more to qualify for the final payment.

But what of his patent application? No doubt Lipperhey was disappointed—albeit probably not surprised—to learn that it had been declined on the grounds that the invention was now well known. Nevertheless, he discharged his obligation to the States General, and on 13 February 1609, delivered the two outstanding binoculars. That day's entry in the States General's account book recording his final payment of 300 guilders is the last we hear of Lipperhey in the historical record until the melancholy notice of his burial in Middelburg on 29 September 1619.

History accords Lipperhey a place as the person who first brought the telescope on to the world stage, although it denies him the honour of being the instrument's original inventor. The uncertain circumstances surrounding the whole affair are sufficient to cast doubt on any one claim to that title. But history grossly underrates Lipperhey in one important aspect of his work—and it is one in which, to this day, his contribution has remained largely unrecognised.

We can be certain from the records of the States General that Hans Lipperhey succeeded in making at least one—and probably three—working binocular telescopes. The overwhelming likelihood is that they were the first two-eyed optical instruments that had ever been attempted. To be sure, they would have been crude, but their very existence speaks of major achievement.

The successful construction of what we would now call a pair of binoculars demands two identical telescopes fixed with their axes parallel, the spacing between them matching the separation of the user's eyes. So difficult is this to achieve that over the next two hundred years, only a handful of instrument-builders even made the attempt. It was not until 1823 that the first commercially successful binoculars were produced, in the form of opera glasses, by Johann Friedrich Voigtländer in Vienna. And the modern binocular, which uses glass prisms to fold up the light-path and correctly orientate the image, did not appear until 1894.

Whether or not he actually invented the telescope, or even understood what he was about, Lipperhey was ahead of his time in the mechanical construction of optical instruments. For that—and for the invention of binoculars—he deserves far more credit than he is accorded.

5

FLOWERING

THE TOUCH OF GENIUS

If history has been grudging in its treatment of Hans Lipper-hey, it has been positively gushing towards the other great name associated with the earliest days of the telescope. But that is not surprising. Galileo Galilei, a professor of mathematics in the University of Padua near Venice, was as remote from the seedy world of small-time spectacle-making as it was possible to get. And it was he who took the Dutchman's little gadget like a baton in a relay race and turned it into an engine of discovery. With it, he rocked the scientific and philosophical world to its very foundations.

Today, the tabloid headlines would blare: 'Forty-five-year-old father of three perfects Telescope—probes Universe'. No doubt the news that he was unmarried and had a mistress would also spice the text. But in 1610, scientific announcements were much more gently made. Galileo himself wrote up his results in a hastily prepared little book that rocketed him to international fame. We know it by its Latin name of *Sidereus nuncius—The Sidereal (or Starry) Messenger*—but its title page in translation reveals more cogently (if a little verbosely) what it was all about:

SIDEREAL MESSENGER
unfolding GREAT and VERY WONDERFUL sights
and displaying to the gaze of everyone,
but especially PHILOSOPHERS and ASTRONOMERS,
the things that were observed by
GALILEO GALILEI,
FLORENTINE PATRICIAN
and public mathematician in the University of Padua,
with the help of a SPYGLASS lately devised by him,
about the face of the Moon, countless fixed stars, the Milky
Way, nebulous stars, but especially about
FOUR PLANETS
flying about the star JUPITER at unequal intervals and
periods with wonderful swiftness; which, unknown by
anyone until this day,
the first author detected recently and decided to name
MEDICEAN STARS.

No fool, this Galileo. By dignifying the four newly discovered moons of Jupiter with the family name of the ruler of his native Florence—the Grand Duke Cosimo II de' Medici— Galileo hoped to secure ducal favour. In particular, he hoped for a court position that would relieve him of the burden of teaching that accompanied his university post.

The ploy worked. On 12 July 1610, four months to the day after the book's publication, Galileo was appointed philosopher and mathematician to the Grand Duke and chief mathematician in the University of Pisa. It was a momentous change in his life. But, in the fragmented land that was seventeenth-century Italy, it also meant a move away from the Venetian

Republic and the protection that his position at Padua had afforded him for the previous eighteen years. In that regard, perhaps, it was a change for the worse.

It would be wrong to imagine Galileo as the first person to point a telescope towards the sky. Back in October 1608, the Siamese Ambassador had said of Lipperhey's first telescope that 'even the stars, which ordinarily are invisible to our sight and our eyes, because of their smallness and the weakness of our sight, can be seen by means of this instrument'. Someone, therefore—we have no idea who—had noted the telescope's ability to enhance the eye's sensitivity to faint light very early in the piece, perhaps even during those last few days before Spinola's departure from The Hague.

As news of the telescope spread rapidly out of Holland, reports of its use in astronomy emerged with increasing frequency. A French diarist, Pierre de l'Estoile, wrote of hearing about the new invention in Paris on 18 November 1608, little more than six weeks after Lipperhey had presented himself to the States General. By the end of April 1609, the same man had reported seeing telescopes 'about a foot long' (30 cm) on sale at a spectacle-maker's near the Pont Marchand. Perhaps it was via Paris that news of the telescope reached England, where it was seized upon by at least one well-to-do astronomer.

This man was Thomas Harriot (c.1560–1621), a famous Elizabethan scientist who had been a tutor to Sir Walter Raleigh. We know from his own records that as early as 26 July 1609, Harriot was observing the Moon from his home near London. On that night, he made the earliest known sketch of the lunar surface using a telescope. It was little more than a few scribbles, but the suggestion of craters confirms that it was drawn with more than just the unaided eye. We can only speculate as to the source of his telescope; it was clearly superior to the ones on sale in Paris since it had a magnification of six times which, with the lenses available at the time, would have necessitated an instrument considerably longer than the 30 cm reported by de l'Estoile. Perhaps it had been developed especially for Harriot by his own instrument-maker, Christopher Tooke.

Harriot had a pupil, Sir William Lower, who also observed the Moon at about the same time. The historian Henry King has described Lower's account of what he saw as 'quite unique'. It is a charitable epithet, given Lower's style:

> A little after [new moon], near the brimme of the gibbous parts towards the upper corner appeare luminous parts like starres; much brighter than the rest; and the whole brimme along looks like unto the description of coasts in the Dutch books of voyages. In the full she appeares like a tart that my cooke made me last weeke; here a vaine of bright stuffe, and there of darke, and so confusedlie all over. I must confesse I can see none of this without my cylinder.

A drawing of the Moon by Thomas Harriot dated 17 July 1610, a year after he had made his first lunar sketch.

One wonders what the cook might have thought of the lovingly made tart being compared with the surface of the Moon.

A little later, in November 1609, the German astronomer Simon Marius saw stars accompanying Jupiter—or, at least, that is what he claimed in his *Mundus jovialis* (*Jovian World*) of 1614. Subsequently, this became an issue of dispute between Marius and Galileo, since it was not until January 1610 that Galileo had begun observing the Jovian satellites and he felt Marius was trying to appropriate his discovery. While it is clear that Marius had not at first recognised the significance of what he had seen, he may indeed have been making more careful observations of Jupiter at about the same time as Galileo. If nothing else, the episode vindicated Galileo's haste in going to press with his discoveries in *Sidereus nuncius*.

It also shows that telescopes good enough to reveal Jupiter's moons were being made in Holland by mid-1609. Marius explains in *Mundus jovialis* that despite the best efforts of his patron (Johann Philip Fuchs von Bimbach) to find someone in Germany who could make suitable lenses, he eventually had to resort to buying a telescope from the Dutch. 'This,' says Marius, 'was in the summer of 1609.'

Once again, the specialised and unorthodox nature of the lenses required by early telescope-makers was highlighted. Even the expert lens-grinders of Nuremburg—the best in Germany—could not produce the goods.

THE STARRY MESSENGER

In *Sidereus nuncius*, Galileo says that he first heard a rumour about the Dutch *perspicillum*, or spyglass, sometime around

May 1609, and that it was confirmed by a letter from a former student, Jaques Badovere, in Paris. There are at least two possible routes by which the rumour might have reached him. The first is via a friend in the Collegio Romano, Christopher Clavius. This renowned scholar was one of four Jesuit mathematicians who used the telescope that had been sent to Rome by Guido Bentivoglio in April 1609 to look at the night sky. The second is by way of another colleague and friend, Paulo Sarpi of Venice, who, by December 1608, was aware of the events that had taken place in The Hague two months earlier. It was, in fact, Sarpi who had written to Badovere in France for confirmation of the rumour.

Either way, Galileo's interest was piqued, and he acted quickly to understand the optical principles on which such an instrument might be based. But not for him the chance juxtaposition of two lenses. As a mathematician, he was familiar with refraction—the bending of light rays when they cross a transparent surface such as the front of a glass lens. (He was not, however, acquainted with the modern laws of refraction. They were arrived at in 1621 by Willebrord Snel of Leiden— yet another of Tycho's former associates—and attributed to him in a publication by René Descartes in 1637. In fact, we now know that Thomas Harriot had deduced them as early as July 1601, but had not published them.)

Galileo quickly discovered the basic principles on which the telescope worked, and probably deduced there and then that the magnification is simply the ratio of the focal lengths of the two lenses. He built a prototype using readily available spectacle-lenses fitted into a lead tube which, he says in *Sidereus nuncius*, magnified three times. A succession of improvements followed as Galileo perfected techniques for grinding and polishing his own lenses. Instruments with magnifications of eight, twenty and eventually thirty times appeared, and it was with the last two that the learned professor's epoch-making observations were made.

Sidereus nuncius positively radiates the excitement Galileo felt as he plucked discovery after discovery from the sky with his new instrument. The little book begins with his investigations of the Moon and stars—revealing, among other things, that the heights of lunar mountains can be calculated from the positions of their sun-drenched peaks in the dark portion of the Moon, and that the Milky Way is made not of celestial milk, but of faint stars. It then goes on to describe in detail the dramatic observations of Jupiter that had led Galileo to conclude that it is accompanied by four satellites. These observations had started on 7 January 1610, and lasted until 2 March—only ten days before the book's publication. Even today, the freshness and immediacy of Galileo's narrative are striking.

In its impartial style, *Sidereus nuncius* has more in common with a modern scientific publication than with most seventeenth-century texts. Compare, for example, Galileo's introductory comments with William Lower's description of the Moon:

> The diameter of the Moon appears as if it were thirty times . . . larger than when observed only with the naked eye. Anyone will then understand with the certainty of the senses that the Moon is by no means endowed with a smooth and polished surface, but is rough and uneven and, just as the face of the Earth itself, crowded everywhere with vast prominences, deep chasms, and convolutions.

Admittedly, Galileo is addressing the readers of a book rather than writing to a friend as Lower was (to Thomas Harriot), but the contrast between the two could hardly be more striking.

The first two of Johannes Kepler's three laws of planetary motion—that the planets move in elliptical paths with the Sun at one focus of the ellipse, and that the Sun–planet line sweeps out equal areas in equal times—had been published in *Astronomia nova* in 1609 (see Chapter 2). But the Sun-centred Copernican theory was still far from widely accepted. In particular, the Roman Catholic Church was vehemently opposed to anything that smacked of Copernicanism, declaring that it was incompatible with the doctrines of Holy Scripture.

Nothing in *Sidereus nuncius* promoted Copernicanism, of course, although the idea of satellites orbiting Jupiter meant that the Earth could no longer be regarded as the centre of all motion. Nevertheless, the book received its imprimatur—the licence granted by Church authorities (including representatives of the Inquisition and the Congregation on Blasphemy) that permitted the book to be published. All was well.

Late in 1610, shortly after Galileo had moved to Florence to take up his new job as mathematician to the Grand Duke, he made another significant discovery with his telescope. The planet Venus exhibits phases just like the Moon—from a narrow crescent to a full, circular disc. Galileo's observations confirmed the suspicion that the planets shine by reflected light from the Sun, and that Mercury and Venus at least must revolve around the Sun. They strengthened his conviction that the other planets, too—Earth included—must orbit the same fiery centre. But such outright Copernicanism was dangerous: only a decade earlier, Giordano Bruno—'the mad priest of the Sun'— had been burned at the stake in Rome's Campo dei Fiori for advocating such heresies.

A new book, published in 1613 and now expressing overtly Copernican sympathies, put Galileo on a collision course with the Church. During a visit to Rome in 1616, he was warned not to defend or hold the basic tenets of Copernicanism, but more books followed. Eventually—inevitably—Galileo was summoned before the Inquisition. In the face of its awesome power, he recanted.

In 1633, having barely escaped with his life, he was imprisoned in his own house for the rest of his days. As blindness overtook him, he devoted himself again to the studies that had occupied him in the years before the telescope. And, by the time of his death on 8 January 1642, he had invented the new science of dynamics. Though Galileo had no means of foreseeing it, his work had paved the way for the genius of Newton to develop his universal theory of gravitation.

TYCHO'S PROTÉGÉ

Unlike the affable Galileo, Johannes Kepler was not an easy person to get on with. He had no time for small talk or everyday pleasantries. He paced around incessantly. He could be terribly rude, reproaching people with heavy sarcasm. And, on top of all that, he had little regard for personal hygiene. Even by the standards of the day, it was hard to ignore the industrial-strength aroma he brought with him.

But Kepler was a truly remarkable man—notwithstanding his distaste for washing. Out of a life that seemed to go from one episode of misery to another—from illness, bereavement and religious persecution to his elderly mother's outrageous trial for witchcraft—he prised some of the most important intellectual advances of the early seventeenth century. He was gifted with

Johannes Kepler, discoverer of the Solar System and improver of the telescope.

extraordinary powers of reasoning and a clarity of abstract thinking that places him among the giants of mathematics. And he was also a pious man, earnestly and sincerely humble before his Maker.

Kepler's relationship with Tycho Brahe—whom he joined in Prague in October 1600 after a preliminary visit earlier in the year—was crucial to his success. To be sure, their early dealings with one another were marred by bitter and open conflict—mostly over minor domestic issues, and as much a consequence of Tycho's imperious nature as Kepler's touchiness. But it was Kepler in whom the great astronomer confided as he lay on his deathbed a year later. And it was Kepler whom he begged to carry out a complete analysis of his planetary observations so his life's work would not have been in vain.

On 26 October 1601, two days after the master's death, Kepler was appointed Imperial Mathematician to Rudolph II. He was 29 years old. He now had both the freedom and the resources—at least for a time—to devote himself to the task of understanding the stately movements of the planets among the stars. Although there were difficulties at first with the issue of who actually owned Tycho's observations, Kepler persevered with the work, eventually arriving at the epoch-making conclusions he set out in *Astronomia nova* in 1609.

But very soon afterwards, in March 1610, his attention was diverted by news of the remarkable discoveries Galileo had made using his new telescope. Kepler corresponded enthusiastically with Galileo, declaring that the discovery of Jupiter's moons in particular was strong support for his own view of the Solar System. Indeed, he went further:

> The conclusion is quite clear [he wrote to Galileo]. Our Moon exists for us on the Earth, not for the other globes. Those four little moons exist for Jupiter, not for us . . . From this line of reason we deduce with the highest degree of probability that Jupiter is inhabited.

Kepler's great (but completely unjustified) leap of imagination probably represents the beginning of modern scientific speculation about the possibility of life beyond the Earth. Galileo's long-delayed response was far more measured, but he did thank Kepler for his unqualified acceptance of the new observations.

Once the excitement of Galileo's news had subsided, Kepler turned his attention to another problem—one that he was uniquely equipped to solve. What was this *perspicillum*, this telescope thing with which Galileo had made his discoveries? How did it work? Could it be improved? Kepler had already studied optics, and he now devoted his theoretical prowess to investigating the passage of light through lenses and combinations of lenses. And his work on the topic was so thorough that it almost made the learned Italian professor's efforts look childish.

Kepler published his results in a little book called *Dioptrice*—a word of his own invention—which appeared in 1611. It was, to all intents and purposes, a handbook of optical instruments. Perhaps Galileo was envious; he never acknowledged the book's existence, and soon stopped corresponding altogether with its gifted author.

The real gem at the heart of *Dioptrice* was Kepler's proposal for a new kind of telescope, one that would overcome an important drawback of the Dutch (or Galilean) type. As we saw in Chapter 3, that instrument consists of a convex objective lens of relatively long focal length and a steeply curved concave eyepiece lens placed just inside the focus of the objective. The eyepiece intercepts the converging rays of light before they can come to a focus, and makes them parallel again. (To be exact, it is the individual pencils of light emerging from the eyepiece that are made up of parallel rays, just as the ones entering the objective are. But the angles between the pencils from different parts of the distant scene have been magnified—which is why the scene looks bigger.)

The Achilles heel of the Galilean is that the portion of any distant scene that is visible through the instrument—the so-called field of view—is limited by the diameter of the objective and is actually very small. And the greater the magnification, the smaller it gets. Galileo's view of the heavens through his 30-times instrument, for example, was akin to looking through a drinking-straw.

But Kepler perceived a way around this. Suppose you take your Galilean telescope and throw away the concave eyepiece.

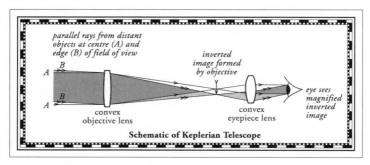

In Kepler's telescope design of 1611, the upside-down image formed by the objective lens is viewed through the simple magnifying glass known as a convex eyepiece.

You are left with a long focal-length convex lens that will project an upside-down image of a distant scene on to a flat surface. (Such a projectable image is described as a 'real' image.) Now imagine the flat surface to be something translucent, like a piece of tracing paper, so you can see the image through it from behind. Then you could take a magnifying glass (another convex lens, this time having a short focal length) and magnify the image on the sheet of paper. And guess what? You'd have a telescope.

As it happens, the translucent screen is unnecessary, because the image can be magnified equally well as it simply hangs in space. That is exactly what Kepler proposed in theorem number 86 of *Dioptrice*: a telescope consisting of two convex lenses of unequal focal length. The objective has a longer focal length than the eyepiece, and the telescope's magnification is just the ratio of the two.

In Kepler's design, the field of view is independent of the diameter of the objective, so it can be made much bigger than in a Galilean. The drinking-straw effect disappears—or, at least, is substantially alleviated. Unfortunately, it is replaced by another drawback. The image seen through the eyepiece is now upside-down. Of course, that is of little consequence when you're looking at stars or planets, but for observations of armies, men o'war or more peaceable terrestrial objects, it is (literally) a pain in the neck. For this reason, Kepler's design is sometimes called the astronomical or inverting telescope. But it is equally well known simply as the Keplerian, immortalising for all time the name of the great—if rather stale—man who invented it.

Unlike Galileo, Kepler was not an experimentalist, so he never built a working version of his telescope. In any case, his eyesight was so poor that he would have been unlikely to wrest any great astronomical discovery from it. Probably the first successful example was built a few years later by his country-man Christoph Scheiner (1573–1650), a Jesuit astronomer who was a noted adversary of Galileo's. And it worked—exactly as Kepler had said it would.

With the publication of *Dioptrice*, Kepler laid aside his work on optics, and returned to his other studies. But bigger events soon overtook him. Political unrest forced the abdication of Rudolph in favour of his brother Matthias on 23 May 1611. Kepler was an unwilling adviser in these affairs, for like Tycho, he had become highly regarded as an astrologer. A year later, following Rudolph's death, he moved to the Danube city of Linz, where he became District Mathematician. Here, he completed his great work on planetary orbits, deriving the third of his laws

Christoph Scheiner, the first to put Kepler's telescope
design into practice.

of planetary motion relating the periods of revolution of the planets about the Sun to their mean distances from it. Kepler's announcement was characteristically low-key; the law was quietly included in his *Harmonices mundi* (*Harmonies of the World*) published in 1618.

Further work enabled him to complete the extensive tables of planetary positions he had begun with Tycho back in 1601. These *Rudolphine Tables* were published in 1627. And—ever faithful to his great mentor—Kepler was still hard at work analysing Tycho's observations when he died, in the imperial city of Regensburg on 15 November 1630. He was 58.

6

EVOLUTION

THE TELESCOPE GOES TO EXTREMES

It is a measure of Kepler's grasp of the theory of optics that in *Dioptrice* he explains how the major drawback of his telescope (its inverted image) can be eliminated. The insertion of a third convex lens, correctly positioned between the other two, will re-invert the image so that the view presented to the eye is the right way up. For this reason such a lens is often called an erecting lens. There is a price to pay, though: the telescope has to be made longer—to be exact, longer by four times the focal length of the erecting lens. That is why ordinary drawtube telescopes (which still use the same basic principle) are so long and unwieldy.

However, back in the 1620s, it was not unwieldiness that prevented such telescopes immediately rendering the Galilean type obsolete. There were other reasons—and the first was a real show-stopper. It was simply that *Dioptrice* had virtually disappeared as soon it had been published. Curiously, while the book was received with enthusiasm in England, on the

Continent it was all but ignored. On top of that, any optician who did happen to know of Kepler's trick for re-inverting the image would have discovered that the introduction of another 1620s-quality lens into the light-path would do more harm than good, because of its imperfections. The image presented to the eye, while decidedly erect, would be unacceptably blurred. For a while, therefore, the Galilean telescope remained state of the art—despite its drinking-straw field of view.

As we have seen, the Galilean was originally unveiled in 1608 with military applications in mind. Surprisingly, though, its acceptance by bearers of arms was far from universal. For example, in November 1614, during a struggle for colonial supremacy between the French and the Portuguese, a sea battle took place at Guaxanduba off the coast of Brazil. During a lull in the fighting, the commander of the Portuguese forces used a telescope to check on the activities of the enemy. But he was taken to task by a high-ranking military observer from Lisbon, accompanying him to supervise the operation. 'Sir,' the official told him stiffly, 'this is not the time to be looking through telescopes, for it will neither lessen our task nor make our enemies fewer.' One can only guess at the commander's response.

It was astronomers rather than generals and admirals who eventually pressured the optical instrument trade into developing the Keplerian, or inverting type. A spate of astronomical discoveries during the first few decades of the seventeenth century had prompted all who considered themselves men of learning to try to get their hands on a telescope. These wonders ranged from the cloud belts of Jupiter (discovered by a Neapolitan astronomer called Francesco Fontana) to the object we know today as the Andromeda Galaxy, whose distance of 2.2 million light-years makes it the furthest body visible to the unaided eye. Simon Marius, having observed this faint and indistinct object for the first time with a telescope in December 1612, described it poetically as 'like a candle shining through horn'.

So that they could see these wonders for themselves—and maybe stake their own claims to immortality by observing new ones—would-be astronomers wanted telescopes with ever-greater magnifications. For the Galilean type, that meant ever-smaller fields of view, and ever more difficulty in pointing the telescope at the thing you were trying to look at. So, from the late 1630s onwards, Keplerian inverting telescopes started to appear. But then—in the early 1640s—there was another breakthrough.

A Capuchin monk by the name of Anton Maria Schyrle de Rheita (1597–1660) reported that with the aid of a newly invented telescope he had discovered more satellites around Jupiter. The details of his telescope remained a closely guarded secret, but in 1645, he published an account that mentioned instruments containing up to *four* convex lenses. With an eye to commercial success, he also mentioned that all his telescopes could be bought from an optician in the city of Augsburg called Johannes Wiesel (1583–1662). If you

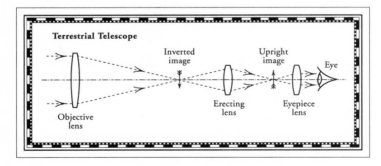

The addition of another convex lens to turn the image right way up transforms Kepler's inverting instrument into a terrestrial telescope.

wanted to know how Rheita's telescope worked, you had to buy one. And it would be expensive, for Wiesel was a top-grade instrument-maker.

The breakthrough here was twofold. First, in the hands of Wiesel and others, lens-making had progressed to the stage where the introduction of additional optical components into the light-path did not hopelessly degrade the image because of manufacturing flaws. And secondly, Rheita had discovered that these additional lenses, if judiciously placed, could be used to make significant improvements. In the end, it turned out that his two- and three-lens instruments were nothing more than re-inventions of Kepler's inverting and erecting telescopes. But the four-lens version was indeed something new. It consisted of an objective, an erecting lens, and an eyepiece made up of two separate components.

These component lenses were what we would refer to today as an eye lens and a field lens. The eye lens did what the eyepiece in a Keplerian telescope had always done: it magnified the real image formed by the objective (or, in this case, by the combination of the objective and the erecting lens). But the field lens, placed slightly nearer the objective lens, had a different function. It provided a further increase in the field of view over the standard Keplerian—hence its name.

By dint of sheer experimentation, Rheita had arrived at a combination of lenses that really did perform better than either Galileo's or Kepler's instruments. And once again, the new invention was greeted with particular enthusiasm across the English Channel. But England was by then a deeply troubled place. Throughout the whole of the British Isles, smouldering political and religious unrest had flared up into revolution. For the last time in its history, the green and pleasant land of England was engulfed in the flames of civil war.

The seven-year conflict between the Royalists of King Charles I (the Cavaliers) and the Parliamentarians (or Round-heads—because they wore the short hair of the working man) culminated in 1649 in the defeat of the Royalists and the establishment of a republic. Under Oliver Cromwell's military junta, the king was put on trial and, on being found guilty of crimes against the state, publicly beheaded. Such was the climate of the day in England. Against this dramatic back-drop it is hardly surprising that a number of prominent astronomers also lost their lives.

Perhaps the most significant was William Gascoigne, who was killed in 1644 at the battle of Marston Moor in northern England, aged only 24. A few years earlier, in more peaceful times, Gascoigne had made an extraordinary discovery. While using a Keplerian telescope, he had seen a spider drop between the objective and the eyepiece, leaving a thread behind it. It happened to be exactly in the focus of the eyepiece, so not only could Gascoigne see the object he was studying through the telescope (in this case—most dangerously—the Sun) but also the magnified spider thread superimposed on it.

Gascoigne realised that by placing two crossed threads in the field of view of his telescope he would be able to point it precisely at a star. Such a telescopic sight would revolutionise the accurate measurement of star positions and, indeed, it became the biggest advance in positional astronomy since the work of Tycho Brahe. Spider threads were perfect for these cross-lines because of their fineness and regularity, but the idea seems to have died with Gascoigne. Subsequent tele-scopic sights used fine wire or silk, and it was not until the second half of the eighteenth century that the idea of using spider threads was rediscovered and quickly adopted. It lasted a long time: as late as the 1960s, the hedgerows around the Royal Greenwich Observatory (by then located in rural Sussex) were being combed regularly for suitable webs for the telescopes.

William Gascoigne also invented a device (again used in the focus of an eyepiece) for measuring the angular diameter of the Sun or Moon and the distances between close pairs of stars. It formed the basis of what became known as the eyepiece micrometer, part of the standard equipment of every major observatory over the next three centuries. One wonders what else this bright young man might have achieved had he survived the bloody battlefield of Marston Moor.

THE STARRIE TUBUS

The Civil War's toll on the noble astronomers of the time included some who survived but fled to the Continent. One such was Sir Charles Cavendish, an acquaintance of Gascoigne's who left England shortly after Marston Moor. Intrigued by the reports of the new multi-lensed telescopes, he sought out Rheita in Antwerp, where the learned monk was supervising the publication of his book on the new discoveries. They obviously got on well, even though their meeting was short:

> We met heere with the famous Cappuchin Rieta [he wrote to a mathematician friend in Britain] . . . I had not discourse enough with him alone to aske manie quaeres, but douteless he is an excellent man and verie courteous, and I found him free and open in his discourse to me.

Sir Charles also enquired about the prices of Wiesel's telescopes, and discovered a wide range of products. By now, telescopes had grown significantly in length from the 5ft 4in (1.6 m) of Galileo's best instrument. True, Wiesel did offer a compact 3 ft (0.9 m) model at the bargain price of 6 ducats, but that was little more than a toy. Most of his serious telescopes were very much longer. A price list of September 1647 shows that his biggest instruments were around 14 ft (or 4.3 m) long. Their prices were 50 ducats for a Galilean, 60 ducats for a Keplerian and no less than 120 ducats for one of the special four-lens terrestrials as used by Rheita. These latter telescopes represented the pinnacle of mid-seventeenth-century optical instrument making.

Despite their extraordinary length, however, they had magnifications not greatly different from Galileo's 30-times telescope. And, like Galileo's, the lenses in them were little bigger than spectacle lenses. So where was the big improvement that thirty-odd years of progress had brought? The answer is that in the Keplerian and terrestrial types at least, the field of view was much larger. But in all of them, there had also been a significant improvement in the quality or distinctness of the images they produced.

This new crispness of image quality had a twofold source. First, as we have seen, the lenses themselves were better than those in Galileo's time—the glass was more homogeneous, the surfaces were more accurate, and there was greater freedom from blemishes. But there was another, more fundamental consideration, and it was related to the long and slender shape of the telescopes.

In fact, the opticians of the time knew that telescope objective lenses were incapable of forming perfect images. Two errors, or aberrations, afflicted the images seen through their telescopes (see 'The aberrant behaviour of lenses' below).

One of them—spherical aberration—was well understood but impossible to correct with the technology of the day, while the other—chromatic aberration—remained a complete mystery. Opticians mistakenly blamed their colourful fuzzy images on spherical aberration, becoming obsessed with trying to make non-spherical surfaces. However, it was eventually demonstrated by Newton that chromatic aberration—the colour error—was by far the larger effect (see Chapter 9).

THE ABERRANT BEHAVIOUR OF LENSES

Since the 1630s, it had been known from the work of René Descartes (1596–1650) that objective lenses whose curved surfaces were shaped like segments of a sphere were incapable of producing perfect images. The focused beam of rays from a star would fail to cross at a single point. Kepler had already hinted at this, but Descartes had the advantage of Willebrord Snel's laws of refraction (1621) governing the bending of light as it crosses a transparent surface. In his *Dioptrique* of 1637, Descartes showed that a spherical surface would always produce a blurring of the image. The phenomenon is still known as spherical aberration—the spherical error.

He also described the surface shape that would correct this effect; it is, in fact, a hyperboloid, which is subtly different from a spherical surface. While a reasonable approximation to a spherical surface could be made in the 1640s, a hyperboloid was way beyond the technology of the day. Sir Charles Cavendish himself, being familiar with Descartes' work, had been involved in attempts to make such lenses before he left England, but he and his colleagues had failed altogether.

Eventually, the obsession with non-spherical lens surfaces became so great that in 1661 the Royal Society set up a special committee to consider how they might be made.

There is, however, another, vastly greater effect whose end product is also a blurring of the images. White light, as it passes through a lens, is dispersed—broken up into its component spectrum colours. The result is that each colour forms an image in a slightly different position from that of its neighbour, with the violet image closest to the lens and the red one furthest away. Any object seen through the eyepiece has coloured fringes around it. The phenomenon is called chromatic aberration—the colour error (see p. 142). Today, we know that both chromatic and spherical aberration can be corrected by using two or more lenses in combination.

In the 1640s, there was only one known way to get around these problems. Making the focal length of an objective very long in comparison with its diameter—so that the curvature of its surfaces would be shallow—reduced both aberrations to a level at which they were not noticeable. The long, spindly telescopes of the day were therefore simply the result of a pragmatic approach to fundamental optical problems. It was little wonder they were sold by the yard. Not surprisingly, though, they brought with them another drawback. The images yielded by the small-diameter objective lenses were faint, so only relatively bright objects could be observed. Nevertheless, before too many years had passed, these long and spindly telescopes were destined to get longer and spindlier still.

It is not known whether Rheita finally talked Sir Charles Cavendish into buying one of Wiesel's telescopes, but we do know that the expensive optician supplied at least two to an English customer. A letter survives in the form of a contemporary copy that is essentially the instruction manual (the 'direction') for the second of these 'starry tubos', which Wiesel completed in December 1649. It makes interesting (and entertaining) reading:

> Honoured Sir . . . My last was that I would deliver the starrie telescopium to Herrn von Stetten & of him recive 100 rix dollars by your ordre, which was this day effected. Though with [the] last tubo a large direction how to fitt the same was sent, [and] that following that said direction one could scarce erre, yet requires also this nightlie starry tubo another direction, which follows.
>
> First there bee eleven pipes or drawers & every one of these pipes [is] marked with 2 letters as A and B. There be fower glases. As before in[to] the small pipe [is] screwed in the great objectivum. In the great leather pipe, there is a shorter tubus with two convex glases screwed in in [sic] the black wood with a strong screw ...

Lengthy details of the assembly follow, after which the new owner is advised how to keep his instrument in tip-top condition:

> Now when all these glases by long usage grow durtie, [then] may the same bee taken out & very well cleansed with the whitest linning & without spott every one layd in againe in due place & screwed up.

Wiesel then goes on to explain that this is the first instrument of a new type, and he is rather proud of it:

> Sir, you may be assured this is the first starrie tubus which I have made of this manner & [it is] so good that it goes farre beyond all others wherof my selfe also doe not little rejoyce. When I looked first throug the same, the moone seemed to my eyes exceeding bigg & not above the length of a foote from my eyes & [I] have found out

others things besides what ever observed & brought to sight. Also I have observed last night being 16 december 1649 in the evening about 7 a clocke Saturnus in such figur as herover . . .

Wiesel's sketch of Saturn has been faithfully transferred to the copy, but it was to be another ten years before the true nature of the planet's rings was recognised. The letter concludes with further advice about how to get the best out of the new telescope, and a postscript suggests that the whole thing might be rested in a shallow trough to keep it from 'bowing throug the length thereof'. With no less than eleven

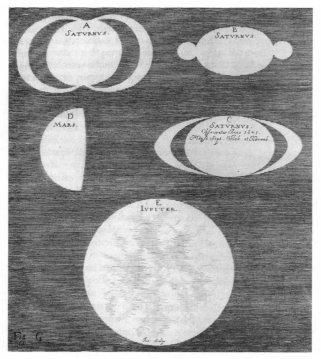

As Hevelius' 1647 sketches of Saturn show, Wiesel was not the only one to be baffled about the nature of the rings.

drawtubes, of which the smallest (rather than the largest) carried the objective, that was probably good advice indeed.

It is clear from Wiesel's description that this 'nightlie starry tubo' is an inverting telescope—a Keplerian type. Yet we are told it contains *four* lenses ('fower glases'), not the usual two or three. Piecing together the details, it seems that these were the objective, the field lens of the eyepiece and an eye lens consisting of two lenses close together—perhaps in contact to give them a very short combined focal length (for high magnification). All were convex lenses, but it is also mentioned that the three lenses constituting the eyepiece were flat on one side (a form known as plano-convex, which was commonly found in early lenses since it was easier to make). Furthermore, Wiesel specifies that all these lenses must have their flat sides towards the eye.

This enterprising man had obviously experimented carefully with the performance of different combinations of lenses in his eyepieces, probably in collaboration with Rheita. While other contemporary opticians had attempted similar tricks (one managing to incorporate as many as *nineteen* convex lenses into a single telescope), it seems that Wiesel came closest to the optimum arrangement for minimising the effects of chromatic aberration in the eyepiece. As it turned out, the person we associate today with the discovery of that arrangement is not Johannes Wiesel, but another great experimenter—the same fellow who finally figured out what Saturn's curious shape was all about. He was a noble and gifted Dutchman who lived from 1629 to 1695, and his name was Christiaan Huygens.

INNOVATIONS

There was great rejoicing in England when the Common-wealth—as the republic was known—finally ran out of steam. Cromwell's puritanical regime had been getting steadily more unpopular—especially since his death in 1659—and it was not long before the House of Commons began to negotiate for the restoration of the monarchy.

> Here out of the window it was a most pleasant sight to see the City from one end to the other with a glory about it . . . and the bells rang everywhere.

So wrote the diarist Samuel Pepys on 21 February 1660, when the end of the Commonwealth was declared. The monarch-in-exile, Charles II, was proclaimed king on 8 May 1660, and crowned in London a year later. In the meantime, just to make sure no one could mistake the sea-change in English politics, Cromwell's body was exhumed and hung from a scaffold on public display.

In London on the day of the coronation, on an extended visit from his home in The Hague, was the accomplished Christiaan Huygens. However, despite the fact that everybody who was anybody had an invitation to the ceremony (and he was certainly somebody), that was not where he was. An event of even greater significance to astronomers was taking place on the same day, a transit of the planet Mercury across the disc of the Sun. So, along with several other men of science, Huygens spent 23 April 1661 diligently observing the transit at the house of an English instrument-maker called Richard Reeve, oblivious to the installation of the popular King Charles.

State-of-the-art telescope of 1647. Within 15 years they had
grown to ridiculous lengths.

Huygens had already made telescopes for himself, the
most impressive being a 23 ft (7 m) long monster with a
magnification of 100 times. It was basically a straightforward
Keplerian instrument with no field lens to its eyepiece, so its
field of view was limited to 17 arcminutes—about half the
diameter of the Moon. While that was much greater than in a
Galilean, there was still more to be gained.

Reeve was making telescopes as long as 36 ft (11 m) on a
commercial basis. Moreover, he had heard of the work of
Wiesel and recognised the advantage of a field lens in the eyepiece. As he had written to a prospective customer in March
1661, one of his telescopes incorporating such a lens would
'take in 40 times more of an object' than the equivalent
Galilean. And that improvement in area was a very valuable
asset to an observer.

No doubt Huygens was impressed with what he saw in Reeve's house. Back in The Hague, he got down to work investigating the possibilities offered by compound eyepieces, and by October 1662 he was able to write to his brother in Paris that he had found something new 'which causes that clarity in the telescopes for daytime [i.e. terrestrial telescopes], and the same in the longer ones [probably astronomical telescopes], giving them at the same time a large field'.

The trick he had discovered was that two plano-convex lenses (a field lens and an eye lens), both with their flat surfaces towards the eye, will, if the ratio of their focal lengths and the separation between them are made to specific values, produce a dramatic improvement in performance. Not only is the field of view much improved, but so also is the quality of the image. It was a major breakthrough in instrument design and, even though the way forward had been paved by Rheita, Wiesel and others, it is known today as the Huygens eyepiece.

When fitted to Huygens' own 23 ft telescope, the new eyepiece doubled the field of view. It gave him great confidence that much longer telescopes were entirely practicable, for an observer would still be able to find his way among the stars using the new device. And he was right. In the end, it was not eyepiece design that limited the lengths telescope-builders would go to.

Curiously, very long telescopes never found much favour in England. Although Reeve and other London telescope-makers like Christopher Cock and the three Johns (John Cox, John Marshall and John Yarwell) produced objectives with focal lengths as long as 60 ft (18.3 m), the emphasis of British astronomy moved in rather different directions.

Under the benign patronage of Charles II, science flourished during the second half of the seventeenth century. The Royal Society was founded in 1660, and household-name scientists abounded. Robert Boyle (1627–1691), Robert Hooke (1635–1703), Isaac Newton (1642–1727) and Edmond Halley (1656–1742) were but a few of the English luminaries who did their respective things at that time. In general, though, astronomy was valued as much for what it could do to enhance national power and prestige as for what it told of the nature of the heavens. Thus, on the Continent, Louis XIV founded the Paris Observatory in 1667 primarily for the improvement of geodesy—the study of the Earth's shape.

On 4 March 1675, in response to the idea that the Moon's motion among the stars might provide a method for determining longitude at sea, Charles II appointed John Flamsteed (1646–1719) as 'our astronomical observator, forthwith to apply himself with the most exact care and diligence to the rectifying of the tables of the motions of the heavens, and the places of the fixed stars . . . for the perfecting of the art of navigation'. Flamsteed, the first Astronomer Royal, was to be paid 100 pounds a year and would have the use of 'a small observatory within our park at Greenwich' which would be designed by the astronomer-architect Christopher Wren (1632–1723).

Unfortunately, no provision was made either for instruments or skilled assistance with which to carry out the accurate positional measurements, and the whole venture might have come to nothing had not Flamsteed dug deep into his own pocket to provide them. His equipment at the outset consisted of two clocks, an iron sextant of 7 ft (2.1 m) radius, a quadrant half as big, and two telescopes of 7 ft and 15 ft (2.1 m and 4.6 m) focal length. Later, he added a Tycho Brahe-style graduated arc, the work of a Yorkshire astronomer and mathematician named Abraham Sharp (1653–1742). Sharp was paid 120 pounds for this instrument, but the price included his lifelong devotion to Flamsteed, for he laboured for years after the great astronomer's death to complete his star catalogue, *Historia coelestis Britannica*.

Thus was the Royal Observatory's inauspicious beginning. Rather more than three centuries later, in 1998, it had an equally inauspicious end when it was closed down by the British Government to free up funds for the nation's participation in a new generation of giant telescopes. Opened by Charles II, it was closed by Tony Blair. Without question, the years between were characterised by great and sustained achievement in astronomy.

DINOSAURS

If the fledgling Royal Observatory did not distinguish itself in advancing the telescope, a handful of individuals on the Continent certainly did. Huygens was one of them. Another was a man called Johannes Hevelius (1611–1687), a brewer from the city of Danzig on the Baltic coast. Like many astronomers of the time, Hevelius was technically an amateur, but his private means allowed him to invest huge sums in the furtherance of his hobby. His first work, published in 1647, was a lavish and highly detailed atlas of the Moon made from his own observations with a 12 ft (3.7 m) telescope.

In the hands of such men, the telescope progressed down an evolutionary path that seems quite futile to us today and inevitably came to a dead end. Nevertheless, along the way it brought new discoveries: more satellites of Jupiter and Saturn, markings on Venus, the rotation period of Mars. And the division in Saturn's ring was observed by Jean Dominique Cassini, an academic at the Paris Observatory, in 1675. All these revelations came from refracting (lens) telescopes that had one thing in common: extraordinary length.

The combination of the high image quality of a long-focus objective and the clarity and wide field of a Huygens eyepiece provided the impetus for a race to build the longest practicable telescope. It gave late seventeenth-century telescope-making a strikingly similar flavour to its counterpart in the early twenty-first century. Both eras are characterised by astronomers driven by the desire to probe ever deeper into the Universe, championing telescopes of massive proportions. And both eras have faced enormous engineering challenges.

By coincidence, even the actual numerical dimensions are similar. But whereas in Huygens' day it was the lengths of telescopes that were measured in tens of metres, today it is their apertures. The CELTs, SELTs and OWLs of this world are as fat as the seventeenth-century telescopes were long.

Hevelius was an early leader in this seventeenth-century super-telescope league. He built a series of instruments that culminated in about 1670 in one whose focal length was no less than 150 ft (46 m). In its workings, this staggering telescope had more in common with the rigging of a sailing ship than an optical instrument. From a mast 90 ft (27 m) high, the telescope's 'tube' was suspended by ropes and pulleys. The tube itself was made up of long planks joined up to make an L-section; at one end was mounted the objective and at the other the eyepiece.

Towering over the rooftops of Danzig, the spindly telescopes of
Hevelius' observatory.

The tube was not light-tight, of course, and could only be
used in darkness. It is easy to imagine the confusion and chaos
that would accompany the operation of such a contraption in
the dark. A large cohort of assistants was required to manipulate the pulleys and point it in the right direction. Any sudden
movement would set the long tube quivering uncontrollably,
and the slightest breeze would make observing impossible.
Against all the odds, though, it could be persuaded to work,
and Hevelius did actually make some observations with it.

The Danzig brewer's ambitions did not stop there. His
book *Machinae coelestis* (1673) depicts a projected astronomical observatory consisting of a central masonry tower supporting up to four long telescopes at a time. It is skirted by a
generous observing platform from which several astronomers
could ply their trade simultaneously, while beneath there is
space for the preparation of yet more of the spindly giants.

Unfortunately, Hevelius' plans were thwarted by a great
fire in 1679, which destroyed most of his observatory. A long,
cathartic letter, written to his patron Louis XIV in Paris
describes the event in revealing detail:

> On that unhappy Evening [before the fire] I felt sorely troubled by
> unwonted Fears in my Soul. To restore my Spirits I persuaded my
> young Spouse, the faithful Aide of my nocturnal Observations,
> to spend the Night in our rural Retreat outside the Walls of the
> City . . .

One hopes that his spirits were indeed restored, for disaster
was around the corner:

> May the Windows of the Human Soul never again look upon such a
> Conflagration as devoured my three Houses with all their precious
> Paintings, Chests of Linen, Wool and Silk, copper and tin Vessels,
> silver Candlesticks, and Ornaments of Gold and Gems . . . the cruel
> Flames have consumed all the [astronomical] Machines and Instru-
> ments conceived by long Study and constructed, alas, at such great
> Cost . . .

There were, however, a few brighter notes:

> . . . if God had not commanded the Wind to turn in its Course, all of
> the Old City of Danzig would surely have burned to the Ground . . .
> Saved by God's Mercy were my Manuscripts, including Kepler's
> immortal Works, which I purchased from his Son, my Catalogue of
> the Stars, my New and Improved celestial Globe, and the thirteen
> volumes of my Correspondence with learned Men and the
> Crowned Heads of all Lands.

Hevelius was in his late sixties, and recognised that he would
have to work hard to recover his losses:

> If such Damage should crush me to the Ground, I whose Locks
> are Hoary and who am not far from my Appointed End, could any
> reasonable Man cast Blame upon me therefor?

Lesser brewers might indeed have been driven to drink, but
Hevelius was made of strong stuff, building new instruments

Hevelius at work with Elisabetha, the 'faithful Aide of my nocturnal Observations'.

and making new observations with them. At the end, however, his advancing years sapped him of the drive and determination needed to take the telescope to yet greater lengths.

Christiaan Huygens likewise produced extremely long telescopes, but he had an even more minimalist approach to their construction. Again there was a tall mast, but it supported only the objective lens fixed in a short metal tube and mounted on a ball-joint, so it could be made to face in any required direction. The eyepiece was held in the fingers or mounted on its own stand; it was connected to the objective by nothing more than a piece of string, which was pulled tight to line up the optical components.

This flimsy arrangement of the bare essentials was euphemistically known as an aerial telescope. The observer's task was supposedly made easier by the provision of a lantern, whose reflection would be seen in the back of the objective when the lenses were properly aligned. However, one imagines that same observer would also have needed large doses of optimism, patience and tenacity in order to succeed.

Three of Huygens' long-focus objectives (made, in fact, by his brother Constantijn in 1686) survive today in the possession of the Royal Society in London. All are about 8 inches (20 cm) in diameter, a significant increase over the spectacle lens-sized objectives of the 1640s. Lens-making had taken giant strides since then, although the quality of the glass itself was still very poor. Their focal lengths were in an even bigger league than their diameters suggest, the shortest being 123 ft (37.5 m) and the longest a cool 210 ft (64 m). How anyone could manage to keep pieces of string taut over these distances beggars the imagination.

Still longer focal-length objectives did exist in the late seventeenth century (up to 600 ft, we are told—nearly one-fifth of a kilometre) but there are no records of them ever being used. With such focal lengths, even die-hard enthusiasts like Huygens and Hevelius must have run out of ideas as to how to observe with them. The use of mirrors to fold up the light-path was suggested by Robert Hooke in 1668, and other ideas were put forward a few years later for keeping a telescope fixed on the ground while light was fed to it by a movable mirror. The problem with all these notions was the sheer difficulty of making acceptably flat mirrors with the existing technology.

From today's perspective, looking back on the era of long telescopes is like looking back to the age of the dinosaurs. We find it astonishing that such creations could ever have existed, and even more so that scientists equipped with such rudimentary technology could have had the patience and perseverance to make them work. The evolutionary trail these instruments followed eventually ran out, and they became extinct—although not, of course, in a cosmic cataclysm like the one that plummeted the dinosaurs into oblivion. Instead, they died out gracefully during the second half of the eighteenth century. By then, technology had moved on. New and better kinds of telescopes had begun to climb the rungs of the evolutionary ladder.

7

ON REFLECTION

BETTER WAYS TO MAKE A TELESCOPE

The world of Islam has had a pretty bad press in recent years. From *fatwa* to *jihad*, from al-Qaeda to Jemaah Islamiah, the media have focused largely on the fanatics and the terrorists. They have paid little more than lip-service to the fact that one-fifth of the world's population lives by this ancient faith in a spirit of peace and harmony, and that Islam was once the single most civilising influence on a troubled world.

A thousand years ago, Islamic culture produced some of the most able thinkers of the day. Of particular relevance to our story is Abu Ali al-Hasan ibn al-Haytham (*c.*965–1039), an Arabic scientist who is better known to scholars by the latinised version of his first name—Alhazen. This early mathematician researched and wrote widely on optics and established such basic principles as the passage of light through space, its reflection from mirrors and its refraction (bending) at transparent surfaces.

Alhazen's work was championed after his death by a Polish

scholar called Vitello, a contemporary of Roger Bacon (see Chapter 3). Astonishingly, both Vitello and Alhazen are immortalised in early English literature, for there is an intriguing reference in around 1387 to a magical mirror (perhaps an imagined telescope) in Geoffrey Chaucer's *Canterbury Tales*. This remarkable instrument could warn of danger from afar or reveal at a distance the infidelity of a cheating husband. Chaucer's characters speculated among themselves on its workings:

> . . . and seyde it might wel be,
> Naturelly, by composicions
> Of angles and of slye reflexions . . .
> They speken of Alocen [Alhazen] and Vitulon [Vitello],
> And Aristotle, that [had] writen in thir lyves
> Of queynte mirours and of perspectives . . .

Chaucer—effectively a senior civil servant in the court of Richard II—was clearly well versed in the sciences, and had not failed to grasp the significance of Alhazen's pioneering studies. It was, however, a sober Latin translation of the Arabic scientist's work published in Basel in 1572 that finally put him on the optical map.

It was probably Alhazen who established that curved mirrors have similar properties to lenses. The focusing effect of a concave (dished) mirror on parallel light from a distant scene, for example, is exactly analogous to the imaging property of a convex lens. Both form an upside-down real image—one that can be projected onto a screen. Of course, there is the rather important difference that the direction of travel of the light-rays is reversed with the mirror, but that did not blind the earliest telescope-makers to the possibilities offered by concave mirrors. If you could use a lens as a telescope objective, why not try a mirror instead?

A flurry of correspondence between a handful of learned men during the first two decades of the telescope's existence explored this intriguing idea. It will probably come as no surprise that the central figure in the discourse was one Galileo Galilei. Letters between Galileo and his friend Gian Francesco Sagredo (1571–1620) discussed a mirror telescope, while Cesare Marsili wrote to Galileo in July 1626 about a mirror telescope reported to have been constructed by another Cesare—Cesare Caravaggi of Bologna. Caravaggi himself was dead by then, but Galileo and Marsili exchanged further correspondence on the matter.

In fact, as it turned out, all these gentlemen had been beaten to it by a Jesuit academic in Rome. Niccoló Zucchi (1586–1670), a professor of mathematics in the Collegio Romano, wrote in his *Optica philosophia* of 1652 that back in 1616 he had tried replacing the convex objective lens of a Galilean telescope with a bronze concave mirror. He had looked into the mirror with the telescope's eyepiece—a concave lens—in the hope of seeing a magnified image of the distant scene behind him. He would quickly have realised that the mirror had to be tilted so that the view wasn't blocked by his own head, but in principle that shouldn't have degraded the image unduly—certainly not beyond recognition.

What Zucchi saw, however, was a blur. Even though his mirror was 'ab experto et accuratissimo artifice elaboratum nactus' (that is, made by a top-notch craftsman), it seems that it was not good enough to produce a satisfactory image. The experiment should have worked in theory—but it failed because of poor optical quality. Given that the craftsman who made the mirror had no doubt produced some first-rate telescope lenses, that must have seemed rather odd to Father Zucchi.

History does not record whether he went back to his crafts-man with some choice words of priestly complaint. But, in fact, the humble optician had no case to answer. What Zucchi didn't realise in 1616 was that it is about four times harder to make an accurate mirror than an accurate lens. And that simple fact held up the construction of a practical reflecting telescope for more than half a century.

At first sight, this peculiarly uncooperative property of a mirror seems like a complete contradiction. Telescope mirrors are made so that the light is reflected from their front surface, and does not pass through the material of the mirror. There-fore, only one surface has to be optically polished, whereas both front and back surfaces of a lens must be polished. Moreover, a lens has to allow light to pass through it, so the glass from which it is made must be reasonably transparent and homogeneous, whereas a mirror need have no such niceties. As it turns out, the issue of homogeneity is a relatively minor one for small lenses (even allowing for the poor glasses

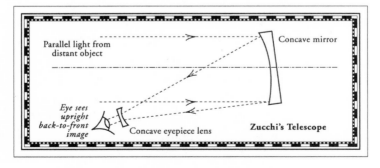

Zucchi's 1616 attempt to make a reflecting telescope used a tilted bronze mirror with a concave Galilean eyepiece. It didn't work.

available in the 1600s), and the root cause of mirrors being so much harder to make is nothing more than the laws of refraction and reflection.

Imagine a ray of light hitting a glass surface that has a slight localised error in the form of a tilt. When the light is refracted (bent) into the glass, its direction is affected by the error—but only at the level of one-third of the tilt of the surface. If it is reflected, however, the effect of the error is *doubled*. Thus, the same inaccuracy in an optical surface produces no less than six times the effect on a reflected ray that it does on a refracted ray.

There is an easily performed demonstration of this subtle phenomenon that anyone can carry out. Half-fill a bathtub with water (you might as well make it warm water so it doesn't go to waste) and compare the reflected image of the upper rim of the bath with the refracted image of the bottom of the bath. Because the refractive power of water is rather less than that of glass, the tilt imparted to refracted rays by surface ripples is now only an eighth of that imparted to reflected rays. Any object on the bottom of the bath remains easy to see when the surface is disturbed, but the reflected image of the bath's side quickly breaks up into meaningless blobs. It is a striking reminder of how hard it is to make a successful telescope mirror. (Of course, once you put a lively four-year-old into the bath, all bets are off.)

In a lens, light refracted at the first surface will proceed through the glass and out at the second surface, where it will have another error superimposed on it because of further inaccuracies in that surface. But even when this second error is taken into account, the performance statistics remain stacked in favour of the lens. There is no escaping the conclusion that the surface accuracy of a mirror must be about four times better than that of the equivalent lens to produce comparable optical performance.

We know that even by the 1640s, optical surfacing techniques were still so poor that lenses bigger than about 20 mm in diameter were useless as telescope objectives. What chance had anyone of producing a telescope mirror using such techniques? Poor old Zucchi. He hadn't a hope of success and had no idea why. Disappointed, and probably not a little disgruntled, he went back to using lenses for his spyglasses. For the time being at least, the idea of a practical reflecting telescope was unceremoniously dumped into the too-hard basket.

TELESCOPES OF THE IMAGINATION

At the time of Zucchi's experiment, the motivation to build a telescope using a mirror as the objective was simply to see whether it could be done. After all, telescopes of any description had been known for only eight years. But as the seventeenth century progressed, the limitations of the refracting (lens) telescope began to make themselves felt. In particular, as we saw in Chapter 6, the mysterious phenomenon of chromatic aberration produced images blurred by spurious colours unless telescopes were made absurdly long and spindly.

It was reasoned—correctly—that this psychedelic effect was caused by the refraction of light as it passed through the lens. With a mirror, however, light simply bounced off the first surface, and therefore did not disperse into an unwanted rainbow of colours. Thus emerged a growing imperative to build a successful reflecting telescope that would side-step chromatic aberration. And it led to one of the most curious paradoxes of early telescope-making.

We have already seen that the refracting telescope progressed throughout the seventeenth century in an empirical and pragmatic way. While its practitioners had little theoretical understanding of what was going on, their continual tinkering with different lenses succeeded in producing acceptable results. On the other hand, mirrors good enough to be used in reflecting telescopes could not be made throughout much of the century. But the *hypothetical* reflecting telescope underwent a period of intense development, entirely in the heads of the leading mathematicians of the day. And this was spectacularly successful. One of today's most eminent telescope designers, Ray Wilson, has noted that the theory of reflecting telescopes was so complete by 1672 that no further improvement was made until the work of Karl Schwarzchild (1873–1916) in 1905.

Who were the players in this extraordinary game of inventing and perfecting virtual telescopes? As a group, they sound curiously familiar: an Englishman, a Scotsman and—well— three Frenchmen. The Irishman appeared much later.

First among them was the great French mathematician, René Descartes, whom we met briefly in Chapter 6. Descartes' scientific interests covered a wide range of problems, and he applied his newly developed technique of analytical geometry to them with great success. Virtually every branch of science still pays homage to Descartes in the use of 'cartesian coordinates'—the representation of positions in space by reference to three mutually perpendicular axes. Less well known is Descartes' lifelong habit of lying in bed every day until 11 a.m., a routine only broken during the last year of his life when his new patron, Queen Christina of Sweden, insisted that he begin work at 5 a.m. In the chilly winter mornings of Stockholm, Descartes promptly contracted pneumonia, and died in February 1650, aged 53.

It is Descartes' work on optics that concerns us, however, and his two books *Le Monde, ou traité de la lumière* (written in 1634 but not published until after his death) and *Dioptrique* (published in 1637) expounded mathematically how the spherical error—spherical aberration—could be eliminated in lenses and mirrors. His work on lenses has already been described (see Chapter 6). Not surprisingly, there was a similar outcome for telescope mirrors. Descartes' theory predicted that a shallow reflecting dish shaped like a segment of a sphere will not produce a perfect image of a star or other distant object. It has to be shaped like a paraboloid, the surface produced when the curved line known as a parabola is rotated about its axis. It is familiar to us today in flashlamp reflectors and satellite dishes. Used in a telescope, a paraboloidal mirror will produce an image free from spherical aberration.

In fact, this much was already well known from less mathematically elegant investigations into the properties of paraboloids. Either way, it made no difference in practice: a useable paraboloidal surface was as unachievable as a spherical one, despite Descartes having turned his attention to the construction of clever new machines to generate non-spherical surfaces. (Such surfaces are known today as 'aspherics'.)

What Descartes' work did achieve, though, was to open the eyes of another French mathematician to the possibilities offered by *combinations* of curved mirrors. A Minorite friar, Marin Mersenne (1588–1648) was in close contact with Descartes on the subject of telescopes. In a book called *l'Harmonie universelle* (1636), Mersenne proposed different kinds of telescopes in which all the optical components were

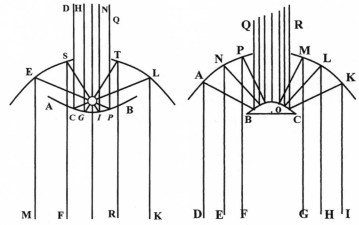

Two of Mersenne's ideas for all-mirror telescopes, from *l'Harmonie universelle*, 1636.

paraboloidal mirrors. As well as offering an all-reflecting solution to the problem of building a telescope, Mersenne neatly solved one of the problems that had dogged Zucchi—namely, how to use a reflecting telescope without your head getting in the way of the mirror.

Imagine a paraboloidal mirror with the dished surface facing a distant object. Parallel rays of light from the object are reflected from the mirror to form a real, inverted image of the distant scene some distance in front of it. (That distance is, in fact, the focal length of the mirror, the image being formed at the focal point.) Now, instead of the light being projected on to a screen at the focal point, imagine that the rays carry on until they hit a second, smaller mirror—also paraboloidal—whose focal point is arranged to be in the same place as the first mirror.

Lo and behold, the rays will be made parallel again as they are reflected from the smaller mirror; moreover, they will be travelling once again in their original direction. Mersenne's breakthrough was to imagine the first mirror to be perforated at its centre so that an observer could look through it into the second mirror. There they would see an upside-down image of the distant scene. If the focal length of the second mirror was shorter than that of the first, the image would be magnified—by an amount equal to the ratio of the two focal lengths.

What Mersenne had invented in this telescope was an exact reflecting analogue of Kepler's inverting telescope described in Chapter 5. Instead of using two convex lenses as objective and eyepiece, it used two concave mirrors whose functions were identical to the lenses. Had it been possible to make one, it would have worked quite well, although its field of view would have been small, partly because there was no optical component equivalent to a Huygens-style field-lens. More importantly, though, it turns out that to get the widest field of view, the eye would have had to be placed *between* the two mirrors, and that is clearly impossible.

Readers who have progressed this far will no doubt spot that there is another possibility offered by Mersenne's thinking. If you can design an all-reflecting equivalent of Kepler's refracting telescope, why not an all-reflecting Galilean-style telescope? Mersenne spotted that, too. One of his other designs envisaged the reflecting equivalent of a convex objective lens with a concave eyepiece lens. They translate into a concave objective mirror exactly as in Mersenne's first design, but now feeding a *convex* 'eyepiece mirror', placed so as to intercept the converging rays from the main mirror before they form an image. Once again, the observer looks through a hole in the main mirror. This instrument would produce an upright image of a distant scene just like a Galilean, but would again have a very small field of view.

Neither of these Mersenne telescopes, incidentally, would produce a view with a dark spot in the middle from the shadow of the second mirror, as might be imagined. Because the eye is focused on infinity when looking through the telescope, the shadow becomes invisible.

Sadly, rather than praising his inventive friend for his new telescope designs, René Descartes wrote to him raising a number of objections. They were not particularly helpful, and the only truly valid one among them was the problem of field size. Descartes seems to have missed altogether the possibilities suggested by Mersenne's designs. These were far-reaching ideas; indeed, compound reflecting telescopes using large concave 'primary' mirrors in combination with a smaller 'secondary' mirror remain today's most important astronomical instruments.

It is possible that Descartes was blinded to these possibilities by the fact that Mersenne's telescopes are 'afocal'—ones in which parallel light goes in at one end and comes out parallel at the other. Of course, that is basically what is required in a telescope for visual use, but more interesting in terms of what might be done with a compound (that is, multi-component) mirror telescope is a 'focal' design. Here, the mirrors combine to form a real image, which can then be magnified using an ordinary lens eyepiece. It was to the Scotsman—a fellow by the name of James Gregory—that the task of inventing the first such optical instrument fell.

8

MIRROR IMAGE

THE REFLECTING TELESCOPE BECOMES
A REALITY

The East Neuk of Fife is a broad promontory of fertile land that separates the estuaries of the Tay and the Forth in south-eastern Scotland. It juts out into the North Sea, that unforgiving tract of water that Billy Connolly once famously described as 'the Arctic Ocean with another name'. On stormy winter nights, the streets of the fishing villages that line its shores echo to the thunder of waves breaking on jagged rocks; on mild summer days, tourists flock to its unspoiled beaches. A 'beggar's mantle fringed with gold'—so James V of Scotland described it in the sixteenth century, and so it remains today, with its patchwork of fields and its golden sands.

In a wide bay on the north-eastern shore of the Neuk lies the Royal Burgh of St Andrews. Steeped in turbulent history, this ancient town is today better known as the home of golf than the site of one of Britain's biggest and most spectacular churches. St Andrews Cathedral was in medieval times as much a magnet for international pilgrims as the Old Course is today. But in June 1559, the Reformation reached St Andrews and the cathedral was sacked, never to be used again for worship. Its broken walls and spires still dominate the town, astonishing visitors with their monumental scale and the lofty vision of their twelfth-century founders.

Associated with the early days of the cathedral was the University—Scotland's first—and that has fared rather better. It continues to flourish in the twenty-first century, not only in attracting royalty to its student body (most recently in the shape of Prince William), but also in attaining very high standards across a broad range of subjects. Among them is mathematics, and it is with the St Andrews School of Mathematics that our story resumes.

When I was a student at St Andrews in the 1960s, the Regius Chair of Mathematics was held by Edward Copson, a genial man best known for his book *The Theory of Functions of a Complex Variable*, which had become a rather unlikely bestseller. Almost exactly 300 years before, the youthful first occupant of that same Chair was entering perhaps the most productive phase of his short life. James Gregory became Professor of Mathematics at St Andrews late in 1668 at the age of 29. This son of an Aberdeenshire clergyman had excelled in mathematics throughout his life, showing

James Gregory, deferential Scots mathematician.

himself to be capable of brilliant and highly innovative scientific thinking.

It is perhaps Gregory's own personality we have to thank for his genius not being more widely known, for in publishing his results he constantly deferred to a younger contemporary whose views he deeply respected—a fellow by the name of Isaac Newton. Nevertheless, among mathematicians, his contributions to such topics as the calculus and series expansions—the bread and butter of the subject—are today widely recognised.

During his six-year stay at St Andrews, Gregory carried out studies as diverse as the determination of longitude from his own astronomical observations, and the discovery that light passing through the feather of a seagull is dispersed into its rainbow colours—a precursor of the diffraction gratings that modern astronomers use instead of glass prisms. Of this discovery he merely wrote, 'I would gladly hear Mr Newton's thoughts of it'—and again failed to gain recognition for a major advance in knowledge.

Gregory was not without resourcefulness, however: when faced with a shortage of funds to build an observatory at St Andrews he went back home to Aberdeen and collected money from parishioners outside the church door. Amazingly, he succeeded and, in July 1673, consulted John Flamsteed as to the types of instruments he should buy. But it is doubtful whether the observatory was finished by the time he left St Andrews in 1674.

Gregory's interest in astronomy went back to his early twenties, when he was still at Aberdeen, and it was here that his work on reflecting telescopes was carried out. He based his studies on the work of Alhazen and Vitello, since he had access to the 1572 translation of their work. Descartes' *Dioptrique* was not available to him, however, and it is remarkable that he arrived at many of the same conclusions from first principles. He wrote up his results in the form of a book, *Optica promota*, and in 1662 travelled to London to oversee its publication.

Thus Gregory began a journey of mathematical discovery, for in February 1663 he went on to Paris, where he was supposed to meet Christiaan Huygens. Unfortunately, the Dutch scientist did not arrive in time, and Gregory had to be content with leaving a copy of the newly published book for him before travelling on to Padua in Italy, where he stayed at the university for almost five years. His return to London in 1668 saw him fêted by the scientific establishment for his work in Italy. He was elected to membership of the Royal Society in June, before moving back to Scotland to take up the new Chair at St Andrews.

What is it about Gregory's work that earns him such a prominent place in our story? The answer lies in the perceptive and innovative study of telescopes he published in *Optica promota*. His careful analysis weighed the relative advantages of three classes of instrument: lens telescopes like those of Galileo, Kepler and Huygens; mirror telescopes like Mersenne's (although he was apparently unaware of Mersenne's work); and a new class of telescopes combining both lenses and mirrors. (For the record, these three types are known as dioptric, catoptric and catadioptric systems respectively—from Greek roots referring to the lenses and mirrors themselves.)

The high point of *Optica promota* is a detailed design in the latter category that has all the characteristics of a modern reflecting telescope. It is known today as the Gregorian design, for Gregory not only designed it, he also made a brave attempt to build one.

To understand the working of the Gregorian telescope, you

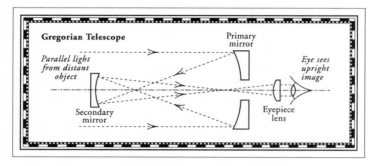

Diagram of a practical Gregorian telescope, showing the concave
primary and secondary mirrors.

have to imagine the Mersenne design of two concave mirrors
facing one another—one large (and perforated at its centre),
the other small. In Mersenne's configuration, they deliver
parallel light when pointed at a distant object, but if their
separation is increased slightly, the parallel light becomes
convergent. It will then form a real image of the distant object
that can be magnified with a lens-type eyepiece. Unlike
Mersenne's design, the eye-point is now accessible behind the
main mirror, and the telescope has a reasonable field of view.
Moreover, it will produce upright images.

Gregory's telescope had all the advantages, including complete freedom from spherical aberration and very little chromatic aberration (since only the eyepiece lens contributes
spurious colour). There was a price to pay, of course, but in
the event it was not particularly significant. Instead of being
paraboloidal, the small secondary mirror had to have a subtly
different shape—an ellipsoid. But since none of these aspheric
surfaces could be generated anyway, it made little difference in
practice.

In order to try out the new design, Gregory commissioned

Richard Reeve to make the requisite mirrors for him while he was in London in 1662. Reeve was much in demand as an instrument-maker, particularly after the transit of Mercury in 1661 (see Chapter 6), but his attempt at producing the mirrors did not meet with immediate success. This was due to the continuing difficulty of producing an accurate reflective surface of *any* kind rather than the fact that the paraboloidal and ellipsoidal shapes could not be made. Gregory had already realised that spherical surfaces would provide a useable approximation, just as they were successfully doing in the refracting telescopes of the day.

With his assistant Christopher Cock, Reeve produced a metal mirror with a focal length of 3 ft (91 cm) as the main concave, together with several smaller secondary mirrors. But the polishing of the main mirror did not meet with Gregory's satisfaction, and Reeve and Cock 'undertook indeed to polish a less speculum to me upon the tool' (that is, make a smaller diameter mirror for him of the same focal length). This Gregory declined, however, 'being about to go abroad; I thought it not worth the pains to trouble myself anie further with it'.

The modern historian Allen Simpson has inferred that Gregory came very close to success, seeing 'but transient views of the object', and it was only because he was in such a hurry to be off on his European travels that he cut the tests short. What is certain is that his mirrors remained in the hands of Richard Reeve, where they eventually found experimental usage in the hands of Robert Hooke. Again, the tests were unsuccessful, but Hooke drew the Royal Society's attention to them in 1672 to show that Newton had not been the first to attempt to make a concave telescope mirror. That was at the start of his long and bitter feud with Newton. In fact, it was Robert Hooke who succeeded in producing the first working Gregorian telescope some two years later.

Gregory, no doubt disappointed with the results of his experiment but elated at the prospect of new mathematical horizons on the Continent, turned his back on his new telescope early in 1663. It was not a permanent separation, however, for almost a decade later, he had to rally to its defence in the face of criticism from his old hero, Isaac Newton.

GENIUS AND SKILL

Richard Reeve, besides being an excellent craftsman, was also something of a trailblazer on the London optical scene. Whereas on the Continent there was a well-established tradition of academic scientists working with optical craftsmen— Rheita had Wiesel, for example, and, in Paris, Cassini had the gifted Italian optician Giuseppe Campani—that was not the case in Britain. However, Reeve had been quietly working with the leading scientists of the day since 1641, when he had attempted to make hyperboloidal lenses for Sir Charles Cavendish (see box in chapter 6). With Reeve's developing skills, and his mentoring of apprentices such as Christopher Cock, there emerged a new breed of professional optical instrument-maker in London. Reeve himself remained pre-eminent, however—so much so that Allen Simpson has described him as the 'English Campani'.

Unfortunately, late in his life, Reeve fell upon seriously troubled times. Somehow—possibly inadvertently—he managed to murder his wife. In a letter to Robert Boyle in 1664, Hooke said:

Perhaps you may have heard of it: if not, in short, he has between chance and anger, killed his wife, who died of a wound she received by a knife flung out of his hand, on Saturday last. The jury found it manslaughter, and he had all his goods seized on; and it is thought it may go hard with him.

Indeed it did, for a while—until his past optical achievements came to the rescue. Not many years before this dreadful event, Reeve had made a long refracting telescope for the newly restored Charles II—who had received it with enthusiasm. There is perhaps no great mystery, therefore, in the fact that some six months after his wife's death, the case of the Crown *vs.* Richard Reeve evaporated in a royal pardon. But Reeve's freedom was short-lived. He died early in 1666—probably a victim of the plague whose morbid fingers touched one in four of the capital's 400 000 inhabitants between 1664 and 1666.

In approaching Reeve for his telescope mirrors, Gregory was acknowledging the new relationship between science and its professional instrument-makers. The fact was, however, that technology had not quite caught up with commercial enterprise, and Gregory's experiment failed. It took something more than mere proficiency to produce the first mirror good enough for a reflecting telescope—something that blended mathematical genius with experimental skills so refined that they eclipsed even a Richard Reeve. It appears there was only one person alive in whom this unique mix of talents lay. And he was—of course—Isaac Newton.

Born on Christmas Day, 1642—the year whose beginning had seen the death of Galileo—Newton was by the late 1660s on the brink of a brilliant academic career. He became Lucasian Professor of Mathematics at Cambridge in 1669, having spent the previous few years distinguishing himself with his as yet unpublished studies of optics. During this period he also formulated his three laws of motion, and laid the foundations of his greatest achievement—the universal law of gravitation. That was eventually embodied in the most far-reaching scientific book ever written, Newton's *Philosophiae naturalis principia mathematica* (*The Mathematical Principles of Natural Philosophy*), published in 1687.

Universally known today simply as the *Principia*, this astonishing work not only explained the motions of planets and satellites as observed by Tycho, Kepler and Galileo, but extended the study to bodies moving through air and water, projectiles shot from guns, pendulums, comets, tides, and the more subtle motions of the Earth and Moon. With consummate insight, Newton's conclusions ranged from the possibility of artificial satellites around the Earth to the idea that all heavenly bodies mutually attract one another. The *Principia* was pivotal in its influence. At a stroke, it solved most of the outstanding problems of astronomy and set the course of fundamental scientific research for the next two hundred years.

In 1666, however, all that was in the future. That was the year in which the youthful Isaac Newton first experimented with the effect of a triangular prism on a beam of sunlight. His deduction that white light is actually composed of a mixture of rainbow colours—christened a 'spectrum'—is one of his

Isaac Newton, gifted builder of the first successful reflecting telescope.

most famous discoveries. The same deduction also led to one
his most celebrated mistakes—that light passing through a
lens will always be decomposed into its spectrum colours, and
so it is impossible to make a lens free from chromatic aberration. As we shall see in Chapter 9, that conclusion was spectacularly wrong, but when such a statement is made with the
authority of a Newton, people tend to believe it. Thus was the
further development of refracting telescopes held up by some
fifty years.

Convinced of the impossibility of making colour-free telescope lenses (a problem he described as 'desperate'), Newton
set to work to see whether he could make a reflecting telescope. As a mathematician, he was quick to spot the fundamental problem in mirror-making described in Chapter 7. A
throwaway line in his *Opticks* of 1704 reveals that he was fully
aware of the problem that had dogged all his predecessors, not
to mention contemporary 'London Artists' such as Reeve:

> For the Errors of reflected Rays, caused by any Inequality of the
> Glass, are about six times greater than the errors of refracted Rays
> caused by the like Inequalities.

To meet this challenge head-on, he first experimented with
materials from which to make his mirror, or 'speculum'. (This
word has a slightly different connotation these days, since it
also refers to a medical instrument for dilating body cavities,
but for two hundred years it was the accepted term for a telescope mirror.) He made an alloy of copper and tin, to which
he added a dash of a whitening agent with an unsavoury
reputation—arsenic. That was to enhance the brittle alloy's
reflectivity and its capacity to take a polish. It was not long
before this and other similar metallic blends were being
referred to as 'speculum metal'.

Next, Newton had to figure out a way of polishing the surface to the required accuracy, and here he seems to have made a real breakthrough. Rather than using a leather or cloth polisher, Newton experimented with pitch, which he melted on to his polishing tool. This allowed the liquid slurry carrying the fine polishing abrasive (Newton used putty powder) to be applied uniformly to the mirror surface, giving an unprecedented degree of control over the amount of metal to be removed—and hence the final shape of the surface. Since the depth of the curve at any point needs to be accurate to about 0.0001 mm, this was a giant leap forward. Pitch polishing is still used today in the optical industry, though usually with jeweller's rouge or cerium oxide as the abrasive.

With these ingredients, and with ample stocks of perseverance, Newton was set to overcome the practical problem of making a telescope mirror. The only remaining question was what layout of optical components he should use for his telescope, and here his native pragmatism shone through. Rather than the complex combinations of curves suggested by Mersenne and Gregory, Newton chose the simplest design he could think of, a concave mirror to produce a real image, with a small flat mirror angled to direct the image out to the side. Here it could be examined with an ordinary lens eyepiece—without the observer's head getting in the way. This arrangement has been called a Newtonian telescope ever since, and has long been the established favourite of amateur telescope-makers because of its simplicity.

Newton's first specimen, made in 1668, was little more than a diminutive working model. As he relates in *Opticks*:

The diameter of the Sphere to which the Metal was ground concave was about 25 English Inches [63 cm], and by consequence, the length of the Instrument was about six Inches and a quarter [16 cm]. The Eye-glass was Plano-convex [flat on one side], and the diameter of the Sphere to which the convex side was ground was about 1/5 of an Inch [5 mm], or a little less, and by consequence it magnified between 30 and 40 times. By another way of measuring I found that it magnified about 35 times.

Note that the mirror's profile was a segment of a sphere, Newton having realised—like Gregory—that this is an acceptable alternative to a paraboloid for a mirror whose diameter is small compared with its focal length. Newton goes on to tell us that the mirror had an 'Aperture of an Inch and a third part'— that is, a diameter of about 34 mm.

Next, Newton pitted his little instrument against a standard Galilean refracting telescope:

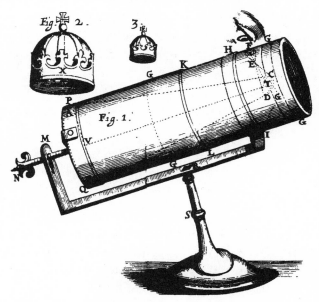

Newton's diagram of his telescope. The tilted flat mirror T allows the image to be viewed from the side of the tube.

> By comparing it with a pretty good Perspective of four Feet [1.2 m] in length, made with a concave Eye-glass, I could read at a greater distance with my own Instrument than with the Glass. Yet Objects appeared much darker in it than in the Glass, and that partly because more Light was lost by Reflexion in the Metal, than by Refraction in the Glass, and partly because my Instrument was overcharged. Had it magnified but 30 or 25 times, it would have made the Object appear more brisk and pleasant.

Therein lay the one weakness of the reflecting telescope in Newton's time—the poor reflectivity of the mirror material made the images look faint. It was a problem not properly solved until the middle of the nineteenth century.

At first, Newton told few people about his new invention. But he made a second example and, on 11 January 1672, presented it to the Royal Society—whereupon they promptly made him a Fellow. Parts of that second instrument still survive, built into what is thought to be an eighteenth-century replica—despite an ancient plaque declaring it to have been 'INUENTED BI SR. ISAAC NEWTON AND MADE WITH HIS OWN HANDS IN THE YEAR 1671'. Of the fate of the first example, nothing is known.

THE THEORY COMPLETED

Thus was the long-sought reflecting telescope brought to reality—six decades after its refracting counterpart. Was Newton the inventor? Or was it Descartes? Or Mersenne? Or Gregory? Newton's great achievement undoubtedly lay in his refinement of optical surfacing technology, but he also approached the design of the instrument with insight and originality. He is, indeed, recognised today as the 'father of the reflecting telescope'. But as we have seen, that epithet has many caveats.

Early in 1672, the gestation of the reflecting telescope took one final turn. A scientist called de Bercé announced to the French Academy that a new form of reflecting telescope had been invented. Its originator was a Monsieur Cassegrain of Chartres. For more than three hundred years, little was known about Cassegrain—not even, with any certainty, his first name. Some authorities suggested he was a Nicolas, and that he was born in 1625 and died in 1712. Others thought he might have been a Jean, or a Guillaume.

Then, in a remarkable piece of detective work completed in 1997, two modern French astronomers (André Baranne and François Launay) uncovered contemporary records proving that he was one Laurent Cassegrain. He was born in Chartres in about 1629, and eventually became a priest and professor in the college there. By 1685, he had moved to Chaudon (Eure-et-Loir), where he died on 31 August 1693.

Cassegrain's invention is singularly important today, for almost all the world's great telescopes utilise exactly this layout. It is, like Gregory's design, a derivative of the Mersenne form using a perforated paraboloidal primary mirror with a curved secondary mirror to feed the light back through the hole. But unlike the Gregorian with its concave secondary mirror, Cassegrain's design utilised a convex secondary to intercept the beam from the main paraboloid *before* it formed an image. It had the effect of refocusing the light back through the hole in the main mirror as a convergent beam, forming a real image that could be magnified with an ordinary lens eyepiece located just behind the mirror.

There are no prizes for guessing that once again the secondary mirror had to have an exotic and—by the standards of the day—entirely unattainable shape. It was, in fact, a hyperboloid. What is more surprising is that almost a decade before de Bercé's announcement of Cassegrain's new telescope, James Gregory had actually attempted to make one. The experimental mirrors that Richard Reeve had produced for him in London included both concave and convex secondaries, revealing that the Scots mathematician had well and truly pre-empted his French colleagues.

With the emergence of Cassegrain's design, the range of possibilities for basic reflecting telescopes was complete, and the theory sufficiently well understood that no further improvement was made for more than two centuries. Again, however, the irony of the situation was clear: even by 1672 only two working examples of reflecting telescopes existed anywhere in the world—those of Isaac Newton.

It was exactly this point that Newton himself made when he responded with some vitriol to the news of Cassegrain's invention:

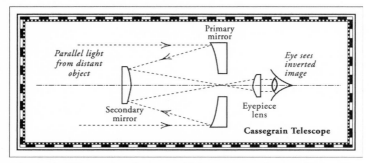

A practical Cassegrain telescope, with its concave primary mirror and convex secondary.

> I could wish therefore M Cassegraine had tryed his designe before
> he divulged it; But if, for further satisfaction, he please hereafter to
> try it, I beleive the successe will informe him, that such projects are
> of little moment till they be put in practise.

Newton's letter to the Royal Society's secretary, Henry Olden-burg, levelled other critical comments at the design, although only that one had any real validity. But he also pointed out that the designs of Cassegrain and Gregory embodied similar principles, and given Gregory's failure to make a working tele-scope, obviously neither of them was any good. Newton's letter was published in the Royal Society's journal, *Philosophical Transactions*.

Up in St Andrews, Gregory heard of this with some dismay, and entered into correspondence with the Royal Society in defence of his invention. The gist of his dispute with Newton was the issue of whether Gregory's experimental telescope of 1662 had failed because of its spherical surfaces (as distinct from the aspheric surfaces it should have had) or because of inaccurate polishing. It was never fully resolved. Sadly, it seems that these two great mathematicians remained ever at cross-purposes.

By now, Newton was recognised as one of the leading scien-tists of his day, and he went on to distinguish himself further in his researches. But his subsequent career was marred by a number of violent disputes with his contemporaries, and these eventually took their toll on his state of mind. In 1693, barely half a dozen years after the *Principia* had been published, he retired from scientific work to become a civil servant. He held prestigious positions at the Royal Mint, first as Warden (1696) and then as Master (1699). In 1705, a knighthood was con-ferred upon him by Queen Anne—the first time such an honour had been bestowed for services to science. Newton died in 1727 at the ripe old age of 84.

No such lengthy twilight years awaited James Gregory. In the early 1670s, he became disillusioned with the attitude at St Andrews to his 'modern' brand of teaching. The University—already more than two and a half centuries old when Gregory was there—was ancient in more ways than one:

> ... the affairs of the Observatory of St Andrews [Gregory wrote] were in such a bad condition, the reason of which was, a prejudice the masters of the University did take at the mathematics, because some of their scholars, finding their courses and dictats opposed by what they had studied in the mathematics, did mock at their masters, and deride some of them publicly.

Gregory left in October 1674 to take up the Chair of Mathematics at Edinburgh. But a year later, at the age of only 36, he suffered a stroke while observing Jupiter's moons with his students. A few days later, he died. Surely, had he lived longer, his genius would have carved out for him a far greater niche than modern history accords him.

There is one final and rather unexpected twist to the early history of the reflecting telescope. As we noted in Chapter 7, the main players in the story were the Englishman (Newton), the Scotsman (Gregory) and the three Frenchmen (Descartes, Mersenne and Cassegrain). These mathematicians made their respective contributions between 1634 (Descartes' *Le Monde*) and 1672 (de Bercé's announcement of the Cassegrain telescope). But there exists an astonishing book, published in Italy in 1632, that actually foreshadows *all* their inventions.

Its author, Bonaventura Francesco Cavalieri (1598–1647),

was a Jesuit monk and self-confessed disciple of Galileo. In 1629, having made a significant contribution to the theory now known as the integral calculus, he was appointed Professor of Mathematics at Bologna. Here, he wrote *Lo specchio ustorio* (*The Burning Mirror*), whose title is a reference to the idea of using mirrors to ignite combustible material from a distance.

Despite its unpromising name, this work is essentially a treatise on the optics of curved mirrors. It explores the imaging properties of paraboloidal mirrors, as Descartes had done. It suggests combinations of mirrors with properties identical to Mersenne's (including the perforated main mirror) and very similar to Gregory's and Cassegrain's. While these are intended as burning mirrors rather than telescopes, the book also explicitly suggests the idea of a reflecting telescope that sounds very suspiciously like Newton's—and that was a decade before Newton was born.

Why was Cavalieri's contribution to the development of the reflecting telescope largely ignored until the historian Piero Ariotti put it firmly on the map in 1975? The answer must lie partly in the broad scope of *Lo specchio ustorio*, which covers not just burning mirrors, but also mirrors to reflect heat and sound.

But Cavalieri may also have unwittingly shot himself in the foot. As he said of reflecting telescopes in *Lo specchio ustorio*:

> I have taken this opportunity to mention such [an idea] but only as something whimsical, to give satisfaction, in other words, to those frivolous people who crave for cake instead of bread. For in my view they will never match the excellence of the refracting telescope either by combinations of mirrors or by the addition of lenses as anyone who wishes to try will, I believe, find out.

Of course, he was wrong. But, on many of the issues concerning reflecting telescopes, so were Descartes and Newton. Bonaventura Cavalieri has been seriously underrated by history. If Newton is to be called the 'father of the reflecting telescope', then surely Cavalieri must be its godfather.

9

SCANDAL

THE TELESCOPE IN COURT

At the turn of the eighteenth century, there were few places in England more unspeakably dreadful than London's Newgate Prison. It is hard for us, as would-be time-travellers from the twenty-first century, to imagine anything like the squalor and violence that held sway behind its grim walls. Though it had been rebuilt after the Great Fire of London in September 1666, nothing about it could be described as enlightened—and there was precious little in the way of comfort for its inmates.

Astonishingly, those unfortunate souls were expected to pay for the privilege of living amid the stench of unwashed bodies and the ever-present threat of typhus. On arrival, an up-front admission fee—'garnish' money—was demanded by other inmates under threat of violence. Then there was the cost of accommodation and food levied by the authorities, the quality of both depending on the prisoner's ability to pay. The well-heeled might seek lodging at the Keeper's House, while residents of the lower ward lay 'upon ragged blankets, amidst unutterable filth . . . the lice crackling under their feet'—as one condemned highwayman vividly described it.

For many inmates, there was only one road out of Newgate, and it led to the gallows at Tyburn. Often, it was for the pettiest of offences. But perhaps even more pitiful was the plight of debtors, who had no resources to feed themselves and were sometimes reduced to living on rats and mice caught in their cells. Their woeful situation was compounded by a classic catch-22. Having served their sentences, all prisoners had to pay a departure fee before they could be released. But debtors, owning nothing, could die in gaol awaiting their overdue freedom. Such was the corrupt and soulless environment of the eighteenth century's correctional service.

It was the spectre of the debtor's prison that haunted several of London's most prominent opticians during the second half of the eighteenth century. Paradoxically, they faced that prospect as a direct result of one of the telescope's greatest advances, a technological leap that had inadvertently brought in its wake acrimony and scandal.

At the root of these unhappy events was an ill-considered judgement by the greatest scientist of the age, Sir Isaac Newton. The science of optics had been transformed in his hands—not only in his studies of refraction (the bending of light) and dispersion (its decomposition into rainbow colours), but also in more subtle phenomena related to the wave-like nature of light. And his contribution to the telescope was not merely the first practical reflector. He had also made a startling discovery about refracting telescopes:

> . . . the greatest Errors arising from the Spherical Figure of the Glass [spherical aberration], would be to the greatest sensible Errors arising from the different Refrangibility of the Rays [chromatic aberration] . . . only as 1 to 1200. And this sufficiently shews that it is not the spherical Figures of Glasses, but the different Refrangibility of the Rays which hinders the perfection of Telescopes.

In other words, the damaging effect of the spherical error in telescope lenses is minute compared with that of the psychedelic colour error (see Chapter 6). At a stroke, Newton had put paid to the long-held notion that all the ills of the refracting telescope could be cured with non-spherical lens surfaces. But—erroneously—he also put paid to any hopes of an alternative solution: 'I do not yet see any other means of improving Telescopes by Refractions alone, than that of increasing their lengths'—which is exactly what all his contemporaries were doing.

It was a nephew of the Scotsman James Gregory who dared to take a reality check and suggest that Newton might have missed something. David Gregory (1659–1708) had been a youthful Professor of Mathematics at the University of Edinburgh, like his uncle James. In 1691, however, he became Savilian Professor of Mathematics at Oxford, an appointment that owed much to the influence of Newton himself. He was also elected a Fellow of the Royal Society in the same year.

David had followed the family tradition of exploring ideas for new telescopes and, in 1695, described his research in a book called *Catoptricae et dioptricae sphericae elementa*. Here, he remarks that the human eye seems to do rather well at forming images unencumbered by spurious colour—and it does it with a lens. He suggests that the eye's complex structure, which involves several different transparent substances, might provide a model for telescope objective lenses:

> Perhaps it would be of Service to make the Object-Lens of a different Medium [from that of other components adjacent to it], as we see done in the Fabric of the Eye . . . in order that the Image may be painted as distinct as possible upon the Bottom [back] of the Eye.

In fact, this idea had not been missed by Newton—in *Opticks* he describes an objective consisting of two glass lenses sandwiching a water filling. But perhaps the great man had already spotted the flaw in the argument—for, indeed, there is one. We now know that the various materials within the eye are not sufficiently different in their refractive powers to provide any significant correction for chromatic aberration. Thus, the image 'painted . . . upon the Bottom of the Eye' is actually rather badly affected by spurious colour, and most of the sorting-out is done not by optics but by the astonishing processing power of the human brain.

Either way, it didn't really matter. The idea of combining lenses of different materials had emerged—even though Gregory's great work was largely ignored by the scientific establishment. And there things rested for some thirty-four years until, in a curious turn of events, the action shifted from the minds of learned mathematicians to the spare-time musings of a London barrister. A *barrister*?

Indeed, he was a rather accomplished lawyer, plying his trade at the Inner Temple—one of England's four Inns of Court. But for recreation, he thought about optics. Chester Moor Hall (d.1767) took his unusual hobby so seriously that he had built a laboratory for optical experiments at his home in Essex. Moreover, he had become obsessed with the idea of making a telescope objective lens that did not suffer from the dreaded chromatic aberration—apparently after reading in Newton's *Opticks* that the great man could see no way in which this might be done.

It is not known whether Hall was familiar with the work of David Gregory, but the solution he proposed in 1729—two years after Newton's death—was essentially the same as the Scottish professor's. Suppose your objective lens was made not as a single piece of glass, but as two component lenses placed one behind the other. If these components were made of different types of glass, and one was convex (thickest in the middle) and the other concave (thinnest in the middle), then maybe—just maybe—their respective colour errors could be made to cancel out. Thus, the image formed by the combination would be free from chromatic aberration—a condition described by opticians as 'achromatic'.

In fact, Newton himself had probably reasoned along similar lines, but his own experiments had led him to believe that different types of glass would spread white light out into its component rainbow colours to exactly the same extent. In other words, all glasses would have equal dispersive power—and that would render the colour correction impossible. In this conclusion, however, he was wrong.

Hall selected for his experiments the two most different types of glass available to him: the regally named crown glass (basically ordinary window glass, named after the blow-and-spin method of producing it), and flint glass, a relatively new invention with the density and brilliance of natural rock crystal. These two materials have sufficiently differing optical characteristics that the combination of a convex crown lens and a concave flint lens with the right optical prescriptions can be made achromatic. That, at least, was Chester Moor Hall's hunch—and he wanted to test it.

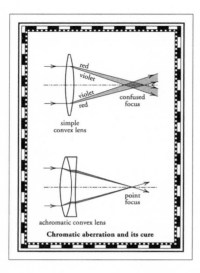

simple
convex lens

achromatic convex lens

Chromatic aberration and its cure

Chromatic aberration and the achromatic lens. By combining different types of glass, the lens can be made to produce images free of false colour.

Hall was not gifted with the skills of a Newton when it came to optical polishing, so he entrusted the manufacture of his lenses to professional craftsmen. But it occurred to him that placing an order for two lenses of suspiciously similar diameter—they were 2.5 inches (63 mm) in the prototype—with the same optician might reveal too much about what he was up to, and could cost him priority in the invention if word got out. So he cleverly ordered one lens from Edward Scarlett in Soho, and the other from James Mann of Ludgate Street—and sat back and waited.

Scarlett and Mann were among London's leading opticians. Scarlett—the more senior—traced his professional lineage back via Christopher Cock to the great Richard Reeve, while Mann was a second-generation optician on the verge of taking on apprentices of his own. Both were busy men with little time for small, one-off jobs like Hall's lenses. It was natural, therefore, that they should both decide to subcontract the work.

What no one could have anticipated—least of all Chester Moor Hall—was that they would choose the same jobbing craftsman to grind and polish the lenses for them. They both sent their work to an optician in Bridewell by the name of George Bass (*c*.1692–1768). It was an extraordinary coincidence—and a fateful one, too. Not only did it eventually place the unfortunate Bass at the centre of a maelstrom of controversy, but ultimately it conspired with other circumstances to rob Hall of his rightful place in the telescope's roll of honour.

SUCCESS AND OBSCURITY

In the events that followed, it is easy to lose sight of the villains and heroes of the piece—and in any case, those roles are defined only indistinctly. But we are fortunate to have the near-contemporary account of another accomplished optician, a fair-minded man who managed to stay above the fray, even though his own personal circumstances threatened at a later stage to draw him deeply into the morass.

This man was Jesse Ramsden (1735–1800), a Halifax-born cloth-worker who had moved to London in his twenties to become an apprentice instrument-maker. He was a great-nephew of Abraham Sharp, whose generous mathematical labours had facilitated the posthumous completion of Flamsteed's *Historia coelestis Britannica* in 1725 (see Chapter 6). Ramsden eventually became the greatest instrument-maker of his time, and is perhaps best remembered today for the design of a telescope eyepiece that rivals Huygens' in its simplicity and effectiveness. It was, however, his innovative work in position-measuring instruments for astronomy, navigation and surveying that won him the Royal Society's highest award, the Copley Medal, in 1795.

In a remarkable manuscript held today in the Library of the Royal Society, Ramsden sets out his understanding of the circumstances following Chester Moor Hall's invention of the achromatic lens in the early 1730s. For yes, it was true—Hall's experiment had worked. The two lenses made by Bass, once returned separately via Edward Scarlett and James Mann to the waiting Hall, produced images spectacularly free from the blurring effect of chromatic aberration when they were fitted together and tested with an eyepiece. Given the travails of telescope makers over the previous hundred years, it was a breakthrough of gigantic proportions.

In his Royal Society document (whose modest title is *Some Observations on the Invention of Achromatic Telescopes*), Ramsden describes Hall as 'a person of extraordinary merit who from a love of retirement and the little thirst he had for public fame; was not as well known to the learned world as he deserved to be'. Ramsden goes on to explain that:

> Chester Moor Hall Esqr. of the County of Essex is the person alluded to, whom I have always been accustom'd to consider as the first Inventor of the Achromatic object Glass. I knew him personally several Years before his Death, which happened about the Year 1767 and he told me that it was from considering the structure of the eye that he first form'd the Idea of what is now called an Achromatic Telescope.

The problem, however, was what Hall chose to do with his invention once he had perfected it by making more trial lenses. Whereas a letter to the Royal Society or an application for a patent might have suggested themselves to a more ambitious person, Hall simply passed the idea on to opticians to whom he thought it might be of some use. After all, he was a lawyer and, having now solved this niggling little optical problem, had no further personal use for it. Ramsden provides us with the details:

> After Mr Hall had satisfied himself of the practicability of his Invention wishing to make it useful, as he himself said he gave written directions for making his telescope to the late Mr Bird. But after 3 or more Years finding Mr Bird too much engaged in making great Instruments to prosecute this business he put the same directions into Mr Ayscough's hands. But his affairs at that time I believe were a little derang'd, he soon after became a Bankrupt.

Thus was the first hint of financial trouble—although at this stage it seems to have been unconnected with the achromatic lens itself. In any event, it appears that two well-known London instrument-makers, John Bird and James Ayscough, failed to recognise the enormous significance of what they had been given—or, at least, they failed to capitalise on it.

Recent circumstantial evidence suggests that a small telescope made around 1735 by James Mann (to whom Ayscough had been apprenticed) might have used an achromatic lens of Hall's design. Ramsden also speaks of a telescope owned by Ayscough that was only 18 inches (46 cm) long, but was capable of revealing the moons of Jupiter. Because of its vastly inferior image quality, an ordinary non-achromatic telescope of the time would need to have been three times that length to perform the same feat. Clearly, with the arrival of the new colour-free lens, the writing was on the wall for those long, spindly, refracting telescopes that were still the mainstay of observational astronomy.

But it was not to be. Not yet, at least. Despite these and other apparent examples of Hall's success, it seems that the impossible happened. Bird and Ayscough failed him completely, and as time went by, the invention of the achromatic objective lens faded—astonishingly—into obscurity.

It was another cloth-worker-turned-optician who rediscovered it—and set the cat among the pigeons of the London optical trade. John Dollond (1706–1761) came from a family of Huguenots (French Protestants) who had fled to England following the annulment of the Edict of Nantes in 1685. Like his father, he was a silk-weaver, but early in his life he developed an interest in optics that rivalled Chester Moor Hall's in its intensity.

Like Hall, too, Dollond maintained his original profession, reserving his work in optics for his spare time. But unlike Hall, he communicated with a wide circle of optical scientists and craftsmen and, throughout the 1730s and 1740s, built up an enviable reputation as an authority on all things optical. At the same time, he and his wife Elizabeth were also raising a family, and when their son Peter (1731–1820) embarked on his own career in 1750, it probably came as no surprise that he chose to become an optician rather than a silk-weaver. What was more surprising was that a couple of years later, his father abandoned weaving with silk to become a weaver of light—and joined him in the business. Thus was the firm of Dollond & Son started, a name that lives on today in the well-known British chain of opticians, Dollond & Aitchison.

The fledgling company soon built up a reputation for innovative and high-quality instruments. Among astronomical customers, it quickly carved out a niche for itself with a new and accurate type of micrometer for measuring small angles, such as the separation of double stars. One of these 'divided object-glass micrometers' (or heliometers) accompanied Captain Cook on his epoch-making voyage to the Pacific in 1769 to observe the transit of Venus—the passage of the planet across the Sun's disc—and fathom the scale of the Solar

System. During that same voyage, Cook also stumbled across the remote coastline he fancifully named 'New South Wales'.

As the 1750s progressed, John Dollond became consumed with the same obsession that had, unbeknown to him, driven Chester Moor Hall a quarter of a century before. How do you solve the problem of making a colour-free objective lens for telescopes? It is a measure of Hall's lack of ambition that despite Dollond's familiarity with contemporary optics and opticians, he had never heard of Hall's invention—nor even of Hall himself.

Dollond's correspondents included some of the Continent's leading mathematicians, men such as Leonhard Euler (1707–1783) of Berlin and Samuel Klingenstierna (1698–1765) of Uppsala in Sweden. Both these scientists were convinced that Newton's assertion concerning the impossibility of making a colour-free lens was wrong. Another sympathetic ear came from Dollond's close friend, the Edinburgh-born optician James Short (1710–1768), who had moved to London in 1738. Short's own enviable reputation came from a single product—his superb Gregorian reflecting telescopes (see Chapter 10)—but he was nonetheless interested in the problem of the achromatic lens. All this discussion only served to fuel John Dollond's consuming passion—until, one day, the inevitable happened.

Once again, it is Jesse Ramsden who provides us with the definitive account:

> Mr Dollond . . . had attempted to make object glasses that would represent objects free from colour; but not succeeding he for the present gave up all hopes of making object glasses with that property. Sometime after this a reading Glass was bespoke of Mr Dollond for the late Duke of York. He applied to Bass, the private working Optician before mention'd, to supply him with one. He shew's him several, and Mr Dollond seem'd to give the preference

to one of flint Glass from its clearness and transparency; but Bass told him that the fault of that Glass was, that letters seen through it towards the edges was much more ting'd with Colours than in one made of Crown glass adding . . . at the same time that he work'd the Concaves for Mr Hall's object Glasses of that glass, that is the flint glass.

One can almost hear the sirens going off in John Dollond's mind as he stood in George Bass's humble workshop in Bridewell. Concave flint glass lenses? Mr Hall's object glasses? Was he really hearing this?

No doubt it was with his head spinning that Dollond headed back to his premises in Exeter Exchange off the Strand. He immediately set about trying to replicate Hall's results, using convex crown glass lenses in combination with flint concaves. More experiments followed, and by 1758 he was sufficiently confident of his analysis that he could write to James Short, setting out his results in some detail. Short, in turn, passed on the letter to the Royal Society, which printed it in *Philosophical Transactions*. At last, the long-sought secret of the achromatic lens was in the public arena.

UNBRIDLED BITTERNESS

But that was when the trouble started. In his letter to Short, Dollond had made no mention of the work of Euler, nor that of Klingenstierna, nor—worst of all—of Chester Moor Hall's earlier success. That gentleman was still living in obscurity in Essex, but Dollond seems to have made no attempt to contact him. If this sounds suspiciously like plagiarism, worse was to follow. While John was the optical genius of the Dollond team, his son Peter was definitely its business brain. It was Peter who

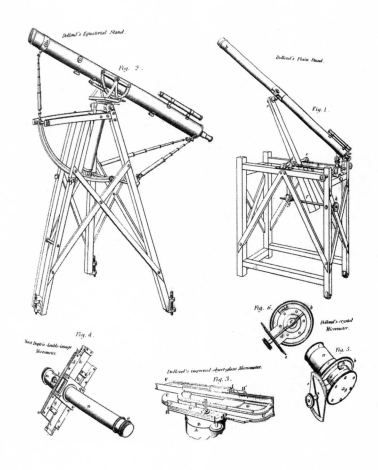

Dollond achromatic telescopes and their accessories.

urged his father to apply for a patent for the achromatic lens and, in April 1758, that patent was granted by King George II. From then on, only the house of Dollond would be permitted to make the fabled doublet lenses.

And make them they did. It is no exaggeration to say that the world beat a pathway to the Dollonds' door. Some of the greatest names in astronomy bought Dollond achromatics as soon as they could lay their hands on them—Nevil Maskelyne of Greenwich (the fifth Astronomer Royal), and Antoine Darquier of Toulouse, for example. Dollond achromatic telescopes also proved a boon to navigators when fitted to maritime sextants. The Royal Society, recognising this outstanding success, quickly awarded the Copley Medal to John Dollond and, in 1761, made him a fellow.

Royal patronage followed: the new monarch, George III (who had succeeded his grandfather in October 1760) appointed him 'Optician to His Majesty'. Eventually, Dollond became the optician of choice of the rich and famous, too, with luminaries such as Thomas Jefferson, Frederick the Great, and an Austrian amateur astronomer called Leopold Mozart (who had a rather gifted son, Wolfgang Amadeus) being named among their customers.

But what of Dollond's competitors? While it might have remained largely unspoken at first, there was almost universal dissatisfaction with the way the patent had been awarded. A handful of opticians ignored it from the word go, risking prosecution for infringement. Information about how to make an achromatic lens was common knowledge, after all, thanks to the publication of John Dollond's detailed studies. But as word of Chester Moor Hall's earlier work got around, the London optical trade began to feel a growing groundswell of resentment.

If there had been any plans to tackle the Dollonds on the issue, they were thwarted by the melancholy news of John's sudden death in November 1761. At the age of only 55, he had suffered a fatal stroke, and was buried at the church of St Martin in the Fields on 6 December—no doubt farewelled by many grieving fellow opticians. Peter became sole partner in the business, which continued to thrive and, once again, to raise the ire of its competitors. Eventually—inevitably—things started to happen.

A man called Francis Watkins, whom Peter and his father had taken on in 1758 to cope with the upsurge in business, was discovered to be making and selling achromatic telescopes privately under his own name. Peter sued for infringement—and won. Moreover, he made it clear that he would take on anyone else who tried the same sort of thing. At that, there was a revolt. Thirty-five London opticians, supported by the Worshipful Company of Spectacle Makers, petitioned the Privy Council in June 1764 to annul the Dollond patent, citing the earlier work of Chester Moor Hall and the fact that achromatic telescopes had been sold in London well before 1758.

Poor old Bass, now in his seventies, was wheeled out in support of the petition. No doubt he testified faithfully about the strange events that had taken place nearly three and a half decades earlier, when orders for two separate but obviously related lenses had arrived in his workshop from Mr Scarlett and Mr Mann; and how, having made the lenses, he had put them together and discovered their purpose, and how he had learned from the two opticians the identity of the idea's originator, a certain Mr Hall. It was an open and shut case.

But it came to nothing. The Privy Council dismissed the appeal, and it was back to business as usual, with Peter taking an even harder line toward his fellow opticians. Whatever your view of the Dollonds' achievements, you have to have some sympathy for those others, for they were doomed to commercial failure if they didn't sell achromatic telescopes, and could expect to be sued if they did. No doubt the spectre of the debtor's prison began to loom large in all their minds.

A series of court actions followed. The test case came in 1766, when Peter sued James Champneys of Cornhill for manufacturing achromatic lenses. Champneys argued strongly along similar lines to the 35 petitioners (amongst whom he had been numbered), but while the court agreed that Hall had indeed been the inventor of the lens, it had some stern words for the defendant. The Chief Justice of the Court of Common Pleas, Lord Camden, noted in his summing up that 'it is not the person who locks his invention in his scrutoire who ought to profit by a patent for such invention, but he who brings it forth for the benefit of the public'. He therefore found against Champneys, who had to pay damages and crippling royalties on his achromatic telescopes. He quickly went bankrupt.

No clearer message could have been sent in support of Peter Dollond, and with that he embarked on a legal spree that saw Addison Smith, Francis Matthews, Henry Pyefinch and—once again—the hapless Francis Watkins bite the dust. These were all men of some eminence, and it is hard to overstate the bitterness that was felt in the optical community at such shocking events. But hard-hearted Peter went from strength to strength. No one was going to challenge the Dollond patent with impunity.

And what of Jesse Ramsden during those terrible days? He had been a noted absentee among the petitioners to the Privy Council. His views are expressed quite clearly in his *Some Observations on the Invention of Achromatic Telescopes*:

The use Mr [John] Dollond made of the hint he received from Bass
is a proof of his great sagacity, and the facts I am stating cannot be
considered as any depreciation of his merit, at least of that merit
which he assumed to himself on this subject, for he always admitted
in his conversation with me that Mr Hall had made the Achromatic
Telescope before him. It is with great pleasure I do justice to the
character of the late Mr Dollond. He was the only man at that time
in London who either had knowledge or a wish to improve Optical
& Mathematical Instruments, and with a most liberal mind as far as
his circumstances would permit, spar'd no expense to improve
both. He instantly saw the application of the two sorts of glass . . .
and being a working Optician it was not long before he brought
these Telescopes into common use.

In essence, that view reflects that of the Court of Common
Pleas at Champneys' trial. There might be some who would
point to the fact that the year before the trial, Jesse Ramsden
had married Sarah Dollond—Peter's sister—as rather a good
reason for him to take a one-sided view of the affair. Certainly,
he stood to gain from patent royalties by his marriage into the
family. But later evidence of tension between Jesse and his
brother-in-law over the issue suggests that Ramsden whole-
heartedly disapproved of the strong line Peter had taken,
notwithstanding his support for the role John Dollond had
played.

There is no greater healer than time in disputes that separate
otherwise like-minded people. And in this instance, it acted
remarkably quickly. Peter showed himself to be almost his
father's equal in the design of colour-free lenses. For, while
an objective lens made from two optical components—a
convex crown and a concave flint—goes a long way towards
being perfectly achromatic, it is only with three lenses that

true perfection can be achieved. Such lenses can withstand very high magnifications and still show bright, crisp images— and Peter made the first of them in 1763.

The modern versions of these triplet lenses use special glasses and sophisticated design techniques; they are called 'apochromats' to distinguish them from ordinary two-element 'achromats'. But in Peter's time, such was the challenge involved in the manufacture of raw optical glass that his lenses had to be relatively small to work at all. His largest three-element telescopes had an aperture (lens diameter) of only 3.75 inches (9.5 cm), but they were the best refracting telescopes of their day. With a focal length of about a metre, they were infinitely more manageable than their long, spindly predecessors. Peter's largest telescope was a 5 inch (12.7 cm) two-element achromat of 10 ft (3 m) focal length, erected at the Royal Observatory in Greenwich.

Grudging admiration for these refined instruments among Peter Dollond's peers eventually led to a softening of their understandably bruised feelings. The Worshipful Company of Spectacle Makers made a gesture of reconciliation as early as 1769 when it elected him to its Court of Assistants. And, by the time John Dollond's controversial patent expired in 1772, Peter had become the centre of a wide circle of optical friends and admirers—just like his father had been twenty years before.

The Dollond company went from strength to strength, popularising small achromatic telescopes with novel collapsing brass tubes instead of the old vellum-covered paper tubes, as well as their larger astronomical instruments. Dollond telescopes were used at the Battle of Trafalgar, and by repute also at Waterloo. Peter himself became Master of the Worshipful Company of Spectacle Makers in 1774, the first of three terms in that office. He continued to live a full and active life until his retirement in 1817 at the age of 86—his sight 'quite worn out in the service of optics'.

He died three years later, leaving behind a dynasty that carried the firm almost into the twentieth century. Few would disagree that the Dollond story provides an unexpectedly happy ending to what had been one of the bitterest episodes in the entire history of the telescope.

10

THE WAY TO HEAVEN

THE REFLECTING TELESCOPE COMES
OF AGE

When Isaac Newton threw down the gauntlet to the mysterious Monsieur Cassegrain in 1672, challenging the Frenchman to put into practice his elegant new design for a reflecting telescope, he knew he was on safe ground. Newton, more than anyone, was intimately familiar with the difficulty of actually making mirrors good enough for a reflecting telescope. Not surprisingly, the gauntlet remained firmly in the dust, and Cassegrain—duly chastised—sank back into obscurity. It is a pity: historians of science had to spend the next three centuries trying to work out who he was.

In 1672, the feisty Newton was the only person to have succeeded in the exacting task of polishing a concave telescope mirror. And, for all he was relatively free with information on his polishing methods, that situation remained essentially unchanged for nearly fifty years. True, his adversary Robert Hooke seems to have succeeded in making a working reflecting telescope to James Gregory's design, which he demonstrated to the Royal Society in 1674. But other attempts—even those made by leading craftsmen in the fledgling optical instrument trade—failed to deliver the goods. In the meantime, long-suffering astronomers were stuck with their ridiculous spindly refractors, for the names of Chester Moor Hall and John Dollond meant nothing in those days.

Then, quite abruptly, the situation changed. The Royal Society's *Journal Book* for its meeting of 12 January 1721 records in characteristically florid style that:

> Mr Hadley was pleased to show the society his reflecting telescope, made according to our President's [Newton's] directions in his *Opticks*, but curiously executed by his own hand, the force of which was such, as to enlarge an object near two hundred times, though the length thereof scarce exceeds six feet [1.8 m], and, having shewn it, he made a present thereof to the Society, who ordered their hearty thanks to be recorded for so valuable a gift.

Suddenly, here was a telescope of the same pattern as Newton's, but of a size that made the great man's handiwork look like a toy. It had a mirror whose diameter—some 6 inches (15 cm)—was the same as the *length* of Newton's telescope. Moreover, when it was tested alongside a refracting telescope 123 ft (37.5 m) long, constructed by the great Christiaan Huygens, it compared most favourably. True, the refractor gave a brighter image, but the mirror telescope offered similar definition and, when it came to convenience and manoeuvrability, won hands down.

As was usual for contemporary aerial telescopes, the Huygens instrument was supported from a tall mast. Hadley's reflector, on the other hand, was mounted on a compact wooden stand

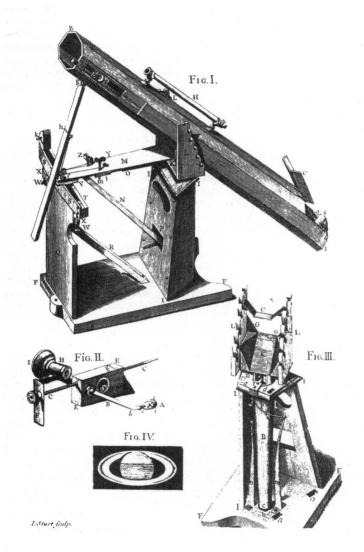

Hadley's reflecting telescope of 1721. With a mirror 6 inches (15 cm) in diameter, it dwarfed the instrument on which it was based—Newton's little telescope of 1668.

equipped with horizontal and vertical motions to allow it to be pointed anywhere in the sky. It seems that the stand was not part of the gift to the Royal Society, however, for soon afterwards, Hadley was 'entreated to lend Dr Halley this apparatus, to have one like it made at the Society's charge, to be used with his noble present'.

Nevertheless, the Royal Society must have been well pleased with the high-performing new telescope. Not only did it boast an excellent concave mirror, but its handsome octagonal wooden tube was fitted with a screw-focusing mechanism and a small 'finder' telescope to make it easy to point the instrument towards objects of interest.

Who was this munificent telescope-maker? John Hadley (1682–1744) was a mathematician and scientific mechanist of some note, who had turned his hand to mirror-grinding in his mid-thirties. His brother George (who is remembered today for his work on atmospheric circulation) worked with him on his first mirror, along with a third brother, Henry. Unlike some other contemporary opticians, John Hadley did not conduct exhaustive tests on suitable alloys for metal mirrors (specula) but he did advance the techniques of mirror grinding and polishing far beyond Newton's (see Chapter 8).

Like the lens-grinders of the previous century, Hadley became obsessed with the need to produce a non-spherical surface for his concave telescope mirrors. The surface must be paraboloidal, departing subtly in shape from a segment of a sphere. Mirror-makers, even today, begin by producing a spherical surface of the right focal length, which they then change to a paraboloid by a process called 'figuring'. It was Hadley who pioneered this technique, introducing the means to reduce local high spots on the mirror's surface with polishing tools he aptly named 'bruisers'.

Moreover, Hadley seems to have been the first person to test by optical means the shape of the surface he had produced. His illuminated pin-hole (located at the centre of curvature of the mirror, where it would conveniently produce an image of itself that could be inspected for flaws) was a forerunner of Foucault's test of 1859, still used by amateur mirror-makers today (see Chapter 15). What Hadley might not have appreciated was that for his first mirror at least, there was actually no need to turn it from a spherical to a paraboloidal surface. With an aperture (diameter) of 6 inches (15 cm) and a focal length of some 5ft 2in (157 cm), the curve on the mirror was so shallow that a simple spherical surface would produce an image good enough for the eye.

Hadley was not one to give detailed accounts of his methods of mirror production, but he did write a description of his figuring process that eventually found its way into a book published in 1738 by one Robert Smith, and entitled *A Compleat System of Opticks*. Though not the most cogent of handbooks, this work turned into a minor bestseller, appearing in both French and German editions. Hadley, for his part, went on to invent the nautical octant, an instrument for measuring the Sun's angular height above the horizon, and the forerunner of the mariner's sextant. His contribution to science was recognised in 1728, when he became Vice-President of the Royal Society.

If Hadley turned the reflector from a toy into a useful astronomical instrument, it was the canny Scot, James Short, who turned it into a commercial success. Like his friend John Dollond, Short had been destined for a career unrelated to telescope-making as a youngster. From the age of 16, his studies at Edinburgh University had led him on a path towards the ministry, a fitting vocation for a bright young man. But in 1731, the 21-year-old student discovered science. It seems to have been as a result of lectures given by the university's Professor of Mathematics, Colin Maclaurin (1698–1746), who was himself something of an expert in telescope design. No doubt that, in its turn, was a legacy of Maclaurin's illustrious predecessor, James Gregory.

Under Maclaurin's guidance, James Short began making reflecting telescopes based on Gregory's pattern. And he turned out to be rather good at it. So much so that by 1734, Maclaurin had declared Short's reflectors to be the best available. One early example, a little instrument hardly more than 16 inches (41 cm) long, was tested in a unique way. A copy of the Royal Society's journal, *Philosophical Transactions*, was propped up some 500 ft (150 m) from the telescope. It could clearly be read by the instrument's owner, a Professor of Scots Law. By a curious coincidence, almost exactly a quarter of a millennium later, I arrived at the same method to test an electronic camera for a large telescope, this time using a page from *The Astrophysical Journal*. Happily, the outcome was the same in both cases—the smiling approval of the authorities of Edinburgh University.

It was not long before Short's proficiency at telescope-making came to the attention of the Royal Society and, in 1736, some of his telescopes were tested by that august body. The result was that James was elected to Fellowship, a remarkable recognition of the young telescope-maker's skills. Having developed a successful technique for manufacturing the two types of concave mirror required in a Gregorian telescope, he milked it for all it was worth, and never looked back. Even before he forsook Edinburgh for the lucrative London market in 1738, Short had amassed a small fortune by building attractive, compact brass telescopes, well mounted on movable stands and yielding the upright images characteristic of a Gregorian. Each one was equipped with a range of eyepieces so that the user could select a magnification appropriate to the observing conditions. To the educated nobility of the time, they were irresistible. To today's antique instrument collectors, they still are.

What was the secret of James Short's success? Primarily, it lay in his figuring method, which, in contrast with Hadley's early work, was absolutely essential because of the steep curves on the Gregorian's paraboloidal and ellipsoidal mirrors (see Chapter 8). Short's technique of imparting these surfaces to discs of speculum metal remained a secret he never disclosed—he is even said to have ordered his tools to be destroyed before his death. Thus, none of his competitors could discover his figuring tricks. He also anticipated modern

James Short's products are exemplified by this beautiful Gregorian telescope, produced in the 1760s.

manufacturing methods by outsourcing the brasswork for his telescopes to other instrument-makers. Though his products were very expensive (double the going rate), they were extremely popular.

By the time of Short's death in 1768, he had produced some 1370 Gregorian telescopes ranging in size up to 18 inches (46 cm) in aperture and 12 ft (3.6 m) long, although most were very much smaller. His instruments had found their way into many of the great observatories of Europe, where they were used for teaching and research. Even though Short himself was a capable astronomer (indeed, he almost became the fifth Astronomer Royal), it is not for his contribution to astronomy that he is remembered. As the twentieth-century historian, Henry King, remarked:

> Short's largest telescopes were few and costly, reserved for aristocratic dilettantes who used them for anything save serious research. For this reason, Short soon found himself in an enviable financial position; for the same reason, his instruments led to few, if any, discoveries of importance.

Which is a great pity.

HEAVENLY MUSICIAN

James Short's lack of scientific ambition for his products contrasts dramatically with the passion of the man who was next to grasp the baton of technological development. 'If not the greatest astronomer of all time, certainly the greatest telescope-maker of all time': thus does one of today's leading telescope experts describe William Herschel (1738–1822), a German-turned-English musician-turned-astronomer. And it is no empty boast. Herschel was the first genuine victim of aperture fever. As the eighteenth century drew to a close, he produced the largest metal-mirrored telescopes the world had ever seen. With them—and with his devoted sister Caroline by his side—he strode through a list of achievements in astronomy that still leaves his biographers gasping in awe.

Wilhelm Friedrich Herschel was born in Hanover, the son of an oboist in the Hanoverian Regimental Guards. He was already an accomplished musician when he moved to London at the age of 19 to pursue his ambitions as a composer. At that time, the musical life of the city was dominated by the work of another German expatriate, George Frideric Handel, who was 72 and totally blind by the time Herschel arrived. The great Baroque composer died only two years later in 1759, by which time the young Herschel had metamorphosed from Wilhelm Friedrich to William Frederick as he immersed himself in the culture of his adopted home.

Herschel's own musical career was very successful, and in 1766 he became organist at the exclusive Octagon Chapel in Bath. Such was his musical influence in that genteel west-of-England town that he eventually took on all kinds of responsibilities—arranging concerts, conducting choirs, giving music lessons and, of course, composing. During the 1760s, he produced a large number of works for orchestra, as well as for small ensembles and solo instruments. Among the latter are his organ pieces, whose buoyant charm can be experienced today in a unique recording made (most fittingly) by a modern astronomer-musician, Dominique Proust of the Meudon Observatory in France.

If the 1760s were idyllic years for William the musician, the 1770s brought a sea-change that was eventually to propel him to world fame. First, though, there was family business to attend to. Moved by compassion for her undeservedly lowly status in the family home, he brought his younger sister Karoline over from Hanover in August 1772 to train as a singer. Just 22, Caroline Lucretia Herschel (as she is remembered today) endured an arduous journey that almost culminated in shipwreck to accompany her brother to his home in Bath. But it was worth it: for the next half-century William and Caroline worked together in an extraordinary partnership that achieved nothing less than an entirely new future for astronomy.

It was soon after Caroline's arrival that the 35-year-old William first felt the pull of the heavens. He had already read widely in mathematics, but in May 1773—as the winter concert season began to wind down—he turned his attention to a book by the Scottish astronomer James Ferguson (1710–1776), a maker of globes and orreries and an able populariser of science. Immediately, he was hooked. He could hardly wait to get his hands on a telescope to see some of these celestial sights with his own eyes. And the celestial sights he had in mind were primarily Solar System objects—planets and comets—for in them lay the essence of astronomy in those days.

William ordered some non-achromatic objective lenses from London and set about building telescopes made of cardboard and tin to utilise them. They were traditional spindly refractors, ranging in length from a relatively manageable 4-footer (1.2 m) with a magnification of 40 to an ungainly 30 ft (9.1 m) monster that defied any attempt to point it in the right direction. Undaunted, Herschel changed tack:

> The great trouble occasioned by such long tubes, which I found almost impossible to manage, induced me to turn my thoughts to

reflectors, and about the 8th September [1773] I hired a 2 feet [61 cm] Gregorian one.

It is tempting to speculate that this instrument was a product of James Short's workshop for, although the Scottish master craftsman had died five years earlier, there were plenty of his telescopes around—and no doubt at prices so far beyond the means of even a well-heeled musician as to preclude him actually buying one.

William was very pleased with the reflector, and determined to make one of his own. Armed with a copy of Smith's *Compleat System of Opticks* and some second-hand grinding tools, he began making telescope mirrors. And 'make' is the appropriate word, for he began by casting his metal specula from an alloy of copper and tin before undertaking the laborious grinding, polishing and figuring that would result in a finished mirror.

Herschel's 7 ft (2.1 metre) telescope, sketched by his close associate, William Watson.

Like James Short before him, William Herschel turned out to be rather good at this. But unlike Short, he quickly discarded the Gregorian design because it was too hard to keep the mirrors in alignment. Instead, he made a 5ft 6in (1.7 m) Newtonian that performed well. Soon, it was followed by an exquisite 7 ft (2.1 m) version. He made several mirrors for this telescope, all about 6 inches (15 cm) in diameter, 'keeping always the best of them for use and working on the rest at leisure'. He would maintain this sensible practice throughout his career.

Meanwhile, Caroline was dismayed to find her new home turning into a cross between a blacksmith's forge and a cabinet-maker's workshop. Not surprisingly, a move to a larger house took place in 1774. The telescopes, too, got bigger, with William making a 10 ft (3 m) reflector having a mirror diameter of 9 inches (23 cm), and beginning work on a 20 ft (6.1 m) reflector, which was completed in July 1776. Another move in 1777 brought some stability—and perhaps some taking stock of the ever-burgeoning telescope situation. William settled on telescopes of 7, 10 and 20 feet (2.1, 3 and 6.1 m) in length for his observing programmes, constantly refining them by making new mirrors. By 1780, the 20 ft had 'a most excellent speculum' 12 inches (30 cm) in diameter.

The mountings for Herschel's telescopes also achieved a new level of refinement. It is fortunate that we have drawings of these early instruments, made by his close friend William Watson (no relation to the author), which record their constructional details. While the 7 ft and 10 ft instruments were arranged on compact wooden stands not unlike Hadley's of sixty years earlier, and the 20 ft boasted a mast and pulley arrangement (based on the even older spindly refractors), all of them now had fine controls for vertical and sideways movement, handily placed near the eyepiece. Convenience and reliability were very high on Herschel's list of requirements for successful observing—and that, after all, was the real motivation for his telescope-building. More than anything else, Herschel wanted to watch the sky—and discover its secrets.

SIMPLY THE BEST

William Herschel started his observing journal on 1 March 1774. Having first satisfied his curiosity as to the appearance of the Moon and planets through a decent-sized telescope, he began a systematic search for double stars—close pairs of stars that he hoped might reveal the elusive 'stellar parallax' (the minute back-and-forth shift in a star's position as the Earth orbits the Sun). That first 'review of the sky', as he called it, began late in 1778, and was conducted using the 7 ft telescope with a magnification of 227 times.

A little more than two years later, on 13 March 1781, the systematic search paid off—but not quite in the way Herschel had expected. He discovered an object that was distinctly disc-like rather than having the point-like appearance of a star; moreover, it appeared to move among the stars as he watched it over the ensuing weeks. At first he thought it might be a comet, but expert opinion disagreed. The expert in question—Nevil Maskelyne, Astronomer Royal—was eventually proved right by the calculations of Anders Johan Lexell (1740–1784), a Swedish astronomer working at St Petersburg. This was no comet that Herschel had found, but a new planet.

It is hard for us to appreciate what a truly momentous event this was. For the first time ever, a planet had actually been discovered. The other five, being visible to the naked eye, had been known since prehistory. Not surprisingly, William Herschel immediately became a household name.

Being a loyal subject, and perhaps having half an eye to royal patronage, Herschel decided to call the planet 'George' in honour of the king. To be exact, he wanted to name it *Georgium Sidus*—the Georgian star—but the idea found no favour among continental astronomers, who had little affection for George III. The name 'Herschel' was suggested by admirers, but the eventual winner was a proposal by the German astronomer Johannes Bode: 'Uranus'. It sounds fine in German, but more than two centuries later, English-speaking schoolchildren still snigger over its anglicised pronunciation—especially now that we know the planet is a gas giant. 'George' would have been far safer.

The aftermath of Herschel's great discovery was, indeed, royal patronage. Not to mention Fellowship of the Royal Society and the award of its Copley Medal. But Herschel's appointment as 'Royal Astronomer' in 1782, and the £200 per year that went with it, were what allowed him to give up music as a profession and devote his efforts to astronomy. This position was brokered largely by William Watson, who recognised that someone trying to be an astronomer all night and a musician all day was bound to run into trouble sooner or later. He had no answer for what followed, however, for Herschel seems to have spent the rest of his life being an astronomer all night and a telescope-maker all day. Fortunately, his telescopes were held in such high regard that he was able to supplement his income significantly by selling them.

Herschel's most important telescopes were those he made for himself. Driven by the desire for more detailed views of the sky, in August 1781 he attempted to cast a mirror no less than

36 inches (90 cm) in diameter for a proposed 30 ft (9.1 m) telescope. The experiment failed, however, when a quarter of a tonne of molten speculum metal poured from its mould on to the stone floor of his basement workshop with explosive violence. Herschel and his workmen were lucky to escape with their lives, and the project was abandoned.

The discouragement did not last for long. Comparisons in mid-1782 of his 7 ft reflector with instruments owned by the Astronomer Royal and a noted amateur astronomer, Alexander Aubert, confirmed that Herschel's telescopes were—to coin a phrase—simply the best. Following another house move, Herschel set about a new project, building what is now called the 'Large Twenty-foot', with a mirror 18.7 inches (47 cm) in diameter. This telescope was finished towards the end of 1783, and proved astonishingly successful.

To start with, it was mounted in a completely new way, with the wooden tube moving vertically between two massive triangular wooden frames some 20 ft (6 m) high, and the entire structure riding on a large turntable. In modern parlance, this is known as an altazimuth mounting, because the telescope moves in *alt*itude (angular height above the horizon) and *azimuth* (horizontal bearing measured from north). Almost all the world's largest telescopes today use a similar arrangement. Herschel's younger brother, Alexander, described it as a 'magnificent and stupendous stand', and the fact that it was blown down by a gale in March 1784 did nothing to deter his enthusiasm. With consummate aplomb, Herschel himself noted in his journal that 'fortunately, it is a cloudy evening so that I shall not lose time to repair the havock that has been made'.

The other innovation in the Large Twenty-foot was that Herschel threw away the small diagonal, flat mirror used in

Herschel's Large Twenty-foot telescope of 1783, forerunner of the Forty-foot.

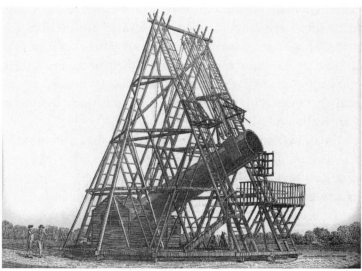

The mirror in Herschel's Forty-foot telescope was a staggering 48 inches (1.2 metres) in diameter.

a Newtonian telescope to bring the light out of the side of the tube. The concave mirror of the Large Twenty-foot was so big that he could simply point his eyepiece at it in the 'front-view' mode that had been tried unsuccessfully by Father Zucchi 170 years earlier (see Chapter 7), tilting the mirror slightly to bring the image to the edge of the tube. A movable platform allowed access to this lofty, downward-looking observing position.

Eliminating the Newtonian flat mirror meant there was only one reflection from a speculum metal surface (which is dull by modern standards), brightening the final image considerably and bringing fainter celestial objects within Herschel's reach. From October 1786, he used only this mode of observing with the Large Twenty-foot, and such an arrangement is still known as a 'Herschelian'.

Herschel's main scientific projects with this telescope and its successors were twofold. First, he wanted to discover the nature of nebulae, those mysterious fuzzy patches of light that are neither stars nor planets, and get their name from their misty appearance. Were they distant clouds of stars, so far away that their light merged into a single diffuse blob? Or was there real 'nebulosity' (mist) out there?

Today, we know that the answer to both these questions is yes—but we now differentiate between many different kinds of objects that look like nebulae. They include galaxies (vast structures of billions of stars at great distances), globular clusters (aggregations of hundreds of thousands of stars orbiting their parent galaxies—including ours) and true nebulae, which are clouds of gas and dust in spiral-type galaxies—again including ours. Herschel had already succeeded in resolving many nebulae into individual stars using his

7 ft telescope, and it was his belief that bigger telescopes would reveal that more, perhaps all, nebulae were made of stars.

Herschel's other passion was what he curiously termed 'gaging'—counting the number of stars in the telescope's field of view in different directions. He was convinced that we ourselves live in a nebula made of stars, and gaging—or gauging—provided a means of estimating its extent in various parts of the sky. These were very far-reaching ideas, and form the basis of our modern picture of the Sun and its family of planets as insignificant members of the vast spiral of stars, gas and dust that we call the Milky Way Galaxy.

Both these scientific projects depended on Herschel being able to detect the faintest possible objects. And he realised that to do that, he needed telescopes of ever-bigger aperture to gather more light. No wonder he was overcome by aperture fever. His aspirations contrast dramatically with the emphasis in the professional observatories of the time, which was on the precision with which the positions of celestial objects could be measured. In that regard, things had changed little since the days of Tycho Brahe—except that the Danish lord's naked-eye sighting devices had now been replaced by exquisite theodolites and quadrants made by gifted instrument-makers such as Jesse Ramsden.

QUANTUM LEAP

It was his thirst for more light that led Herschel to build the telescope for which he is perhaps best remembered—even though it was only partially successful. The 'Forty-foot' was a giant whose 12.2 m long tube housed a mirror no less than 48 inches (1.2 m) in diameter. This famous instrument, whose likeness still adorns the seal of the Royal Astronomical Society, was conceived soon after the Large Twenty-foot was completed.

By the summer of 1785, Herschel had written to Sir Joseph Banks at the Royal Society expressing his concerns about the likely cost of the project, and wondering if the King might be prepared to chip in with a contribution. In an early forerunner of national funding for scientific infrastructure, George III eventually provided two separate grants of £2000 and an annual operating allowance of £200 for the new telescope. These were fairly generous amounts, although in a letter to her nephew John Herschel in 1827, Caroline complained that: 'I never felt satisfyed with the support your Father received towards his undertakings [with the Forty-foot]; and far less with the ingratious manner in which it was granted'. For this, and the wearing effect it had on her brother, she laid the blame squarely on the King's 'shaby mean spirited advisers'.

Herschel pressed ahead with preparations for making two concave mirrors for his new telescope, following his usual practice of duplication. The first mirror was cast in October 1785 at a foundry in London. It did not turn out as well as he had expected, thanks to a 'mismanagement of the person who cast it'. It was thinner than intended, but grinding and polishing went ahead anyway, and it eventually performed tolerably well. Herschel was disappointed that its weight (almost half a tonne) prevented him from exercising his own skills as an optical craftsman. The inverted mirror had to be moved over a stationary tool during polishing and, in the event, needed ten men to do the job. The second mirror, cast in February 1788, weighed twice as much as the first and took as many as 22 men to manipulate it. It was this experience that led Herschel to develop machines for optical polishing, equipment he used exclusively for his mirrors after about 1789.

In the meantime, construction of the instrument itself had begun now that Herschel had found a site for it. During the spring of 1786, he had moved to a house with suitably large grounds at Slough, near Windsor. There he began erecting an enlarged version of the mounting for the Large Twenty-foot. The triangular framework supporting the telescope was more than 15 m (49 ft) high, and was equipped with refinements such as graduated dials and a speaking pipe to communicate with Caroline in the lighted cabin on the turntable where she would make her notes (see 'Observing with William Herschel', below). Once again, Herschel employed the 'front-view' optical layout, observing from a high, movable gallery.

OBSERVING WITH WILLIAM HERSCHEL

Today's astronomers use observational techniques that owe their origin to pioneers like Herschel. But whereas today's astronomers operate their instruments from a comfortable control-room, things were very different in Herschel's day.

Procedure   Herschel systematically scanned the entire sky by eye, 'sweeping' in regular patterns and dictating his observations to Caroline, who made notes by lamplight.

Equipment   Apart from the telescopes and their ancillaries, Herschel used a wide range of eyepieces giving magnifications often measured in thousands. Some were no more than glass beads 0.6 mm in diameter, mounted in tubes of brass or cocos wood. He had a black hood to cover his head and preserve his dark-adapted vision, while Caroline had an accurate clock to note the time.

Observing conditions   Bleak. For example, on the night of 1 January 1783, Herschel's ink froze and his best speculum for the Small Twenty-foot telescope cracked in two. The temperature was −12°C.

Health and safety   Neglected. Observing with Herschel's large telescopes was a hazardous business. Caroline had a nasty fall onto an iron hook hidden under a covering of snow, while an eminent astronomer visiting from Sicily tripped and broke his leg. William himself had several near-misses with collapsing structures, including a beam supporting one of the mirrors for the Forty-foot telescope on 22 September 1807.

The main tube for the Forty-foot was made of sheet iron, and was almost 1.5 m (5 ft) in diameter. While it lay on the ground awaiting its mirror, it proved an irresistible attraction for visitors, who absorbed precious hours of Herschel's time as he explained the workings of the instrument to them. The worst offender seems to have been the King. His proximity at Windsor might have been one of the drawbacks of Herschel's new home, and on one famous occasion in August 1787, he led the Archbishop of Canterbury through the tube. It was Caroline who reported that he remarked genially, 'Come, my lord bishop, I will show you the way to Heaven.' Clearly, George III was not without wit.

Herschel was using the Forty-foot by 1789, but despite the evident increase in light-gathering power and the improvement in the visibility of nebulae, it was not the success he had hoped it would be. The mirrors, having a high copper content, tarnished rapidly and required frequent repolishing (see

'Looking after the Forty-foot', page 178). Moreover, the telescope was impossible to manage without the assistance of two workmen besides Caroline. Herschel also discovered a problem that went on to plague British astronomers until they started building their telescopes on better sites overseas—the weather. Seldom was it good enough for the full 1.2 m aperture to be used. It was far simpler, quicker, and almost as satisfactory to use the Large Twenty-foot. By the turn of the new century, the Forty-foot was getting only infrequent use, and the last observation with it (of Herschel's favourite object, Saturn) took place in August 1814.

Despite the lack of regular usage, Herschel endeavoured to keep the Forty-foot telescope in working order, and the historian Michael Hoskin has suggested that this might have had more to do with impressing the King's guests when they visited than with making useful observations. After all, the King had paid for the instrument, and no doubt wanted his pound of flesh from his Royal Astronomer.

Later in his life, Herschel made a 25 ft (7.6 m) telescope with a 24 inch (61 cm) aperture mirror for the King of Spain—which he completed in 1797 and delivered in 1802. He also built a stubby 10 ft (3 m) for his own use in 1799; this, too, was fitted with a 24 inch mirror. But it is for the Forty-foot that Herschel is best remembered, and that is fitting, given the quantum leap in aperture that it represented. To take the reflecting telescope from a mirror diameter of a few inches to 4 feet within a decade is truly an astonishing achievement.

Perhaps no less significant is the remarkable relationship that existed between Herschel and his sister. Although he had a wife, Mary (whom he married in 1788), and a son, John (1792–1871), who became a great astronomer in his own right, carrying his father's work into the southern hemisphere, it was Caroline who was his astronomical protégée and faithful assistant. She, too, went on to become a capable astronomer, discovering several nebulae and eight comets.

LOOKING AFTER THE FORTY-FOOT

The annual expenses Herschel incurred in repolishing the two mirrors for the Forty-foot are set out in a letter he wrote to Sir Joseph Banks in 1790. It reveals much about the processes involved:

Polishing	£	s.	d.
12 men 6 weeks at 10/6 each per week	37	16	0
Beer for the men, one pint per day	3	12	0
12 polishing dresses and caps washing once each fortnight		19	6
1½ dozen of towels per day washing	1	7	0
1 dozen new for wearing out		9	4

Smith's work in calcining colcothar [jeweller's rouge]. Coals and materials, boiling & calcining vessels. Pounding & preparation of colcothar. Copperas [ferrous sulphate]. 50 lb brown colcothar from the shops. Pales [sic], tubs, ropes, tackle, oil, spirits of wine, pitch, rosin, tar, tar-boiling vessels, brooms, brushes. Camels hair brushes, glasses, pans, carpenter & smiths work & materials.

	15	15	0
Total	59	18	10

When William Herschel died on 25 August 1822, he left behind a remarkable canon of discovery. It includes the planet Uranus and two of its moons, two moons of Saturn, some 1000 double stars, 2000 nebulae and star clusters, and the realisation that the Milky Way represents the plane of a disc-shaped aggregation of stars of which the Sun is a member. On top of that, he was the first person to recognise the existence of 'invisible light' beyond the red end of the spectrum—infrared radiation.

He also left behind a devastated sister. Although Caroline had many friends in England (including the Dollond family), she lost no time in moving back to Hanover, arriving within a few weeks of William's death. There, she continued to receive honours for her astronomical work. She was not, however, the same Caroline. Her frustration shows in a poignant letter she wrote towards the end of her long life:

> You will see what a solitary and useless life I have led these 17 years all owing to not finding Hanover, nor anyone in it, like what I left, when the best of brothers took me with him to England in August, 1772.

She died in 1848, two years short of her 100th birthday.

11

ASTRONOMERS BEHAVING BADLY

MIXED FORTUNES FOR THE TELESCOPE

By the turn of the nineteenth century, on the eve of its 200th birthday the telescope seemed assured of a bright future. The thorny problem of chromatic aberration in refracting telescopes had been solved—albeit amidst a flurry of acrimonious legal wrangling—and long, spindly refractors yielding images fringed with spurious colour were now a thing of the past. The reflecting telescope, too, had progressed by leaps and bounds. While there were those who felt that William Herschel had overreached himself in the construction of his Forty-foot, no one doubted that metal-mirrored instruments could be made successfully with apertures up to 24 inches (61 cm). Such telescopes were giants compared with the previous generation.

As the new century dawned, the two basic telescope types—refractors and reflectors—stood poised on the starting blocks in a race for further development. Each had its strengths and weaknesses and, as the century progressed, vied with one another for superiority. The eventual outcome was settled by the march of technology in this most industrially revolutionised of centuries, but along the way, astronomers continued to prise secrets from the sky with whatever instruments were at their disposal.

Perhaps it was the pressure of competition between supporters of the two types that led to a number of controversial episodes during the nineteenth century. Astronomers, being human, are capable of all kinds of behaviour—including the bad kind. Opticians are the same. We have already seen numerous examples of less-than-wholesome conduct in earlier times (not to mention certain of today's improprieties), but the nineteenth century was particularly well endowed with stubborn and outspoken individuals. Sometimes their behaviour had quite serious consequences for the development of the telescope, but in other instances it merely provided a diversion for their contemporaries—and light relief for today's historians.

Take, for example, the case of Andrew Barclay (1814–1900), a latecomer to the nineteenth-century miscreant league but a spectacularly successful member of it. Barclay was neither an astronomer nor an optician by profession, but a locomotive-builder. In the 1870s, his colliery and shunting engines were the pride of Kilmarnock, the west of Scotland town in which his factory employed more than 400 people.

Sadly, Barclay's mechanical ingenuity was not matched by business acumen, and by 1882 his company was in financial trouble. The problems dragged on until 1893, when he was dismissed from what had now become a limited company. Piecing together the evidence, it appears that Barclay had been guilty of fiddling while Kilmarnock burned for, since the 1850s, he had been tinkering with the manufacture of astronomical telescopes, as an unlikely (and unprofitable) sideline. The few instruments he had produced were, by the standards of the time, small and old-fashioned—Gregorian reflectors with speculum metal mirrors up to 14.5 inches (37 cm) in diameter. They were, however, beautifully engineered, for Barclay had all the resources of his locomotive workshops at his disposal.

Apart from fiscal neglect, two other problems beset Barclay's career as a telescope-builder. The first was that his telescopes were optically hopeless, and the second was that he stubbornly refused to believe that they were. In 1893, he published the first of a number of articles in *The English Mechanic and World of Science* based on observations he had made with his telescopes. Under the heading 'The Unrevealed Wonders of the Heavens', Barclay described such absurdities as egg-shaped protuberances from the planet Jupiter and a Saturn-like ring around Mars—which had also grown a blue, spherical-looking mountain in its southern hemisphere.

These observations produced an immediate and scornful response from the readers of *The English Mechanic*.

> I can only say [wrote one contributor] that if I had a Gregorian telescope . . . that exhibited the great planet as depicted in Mr B's Fig.1, I would dispose of the optical part for what it would fetch, and convert the tube into a chimney cowl straightway.

Another correspondent gently suggested that faulty optical polishing might have led to distortions in the image seen through the telescope. Barclay, however, ignored this lifeline, pressing ahead with reports of brown smoke issuing from Jupiter's mountains and a sketch of Saturn that made the ball of the planet look for all the world like a half-eaten apple.

QVADRANS VOLVBILIS
AZIMVTHALIS

The Revolving Azimuth Quadrant of 1586, one of Tycho Brahe's larger instruments at Stjerneborg. *(The Royal Library, Denmark)*

Where the master walked—the Quadrant crypt today. *(Author)*

State of the art, 1670s. A very early binocular, made by Chérubin d'Orléans
and presented to the Grand Duke Cosimo III de' Medici.
(Istituto e Museo di Storia della Scienza)

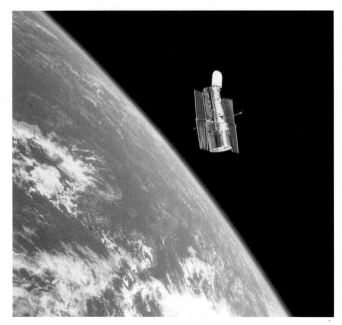

State of the art, 1990s. The Hubble Space Telescope glides over the Pacific
Ocean on the daytime side of its orbit. *(NASA)*

Optical workshop, 1640s. Johannes Hevelius' lens-making tools shown in detail in a charming plate from *Selenographia*,1647. *(Crawford Collection, Royal Observatory Edinburgh)*

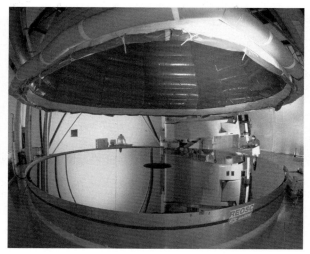

Optical workshop, *c*.2000. One of the 8.2 metre diameter mirrors for the Very Large Telescope in its final stages of manufacture in 1999. *(SAGEM Group and the European Southern Observatory)*

Star trails encircle the dome of the 3.9 metre Anglo-Australian Telescope
at Siding Spring Observatory, near Coonabarabran. They trace the rotation
of the Earth during the 10 hour photograph. Irregular lines of light are
the flashlamps of observers checking the weather.
(*Anglo-Australian Observatory/David Malin Images*)

Seen through a sprinkling of nearby stars, the spiral galaxy NGC 2997 hangs like a radiant jewel in the depths of the Universe. It is a near-twin of our own Milky Way Galaxy.
(*Anglo-Australian Observatory/David Malin Images*)

Still breathtaking after more than half a century, the 200-inch (5.1 metre)
Hale Telescope was for 28 years the world's largest.
(Dr Thomas Jarrett, IPAC/Caltech)

The traditional dome of the 1.2 metre UK Schmidt Telescope at Siding Spring
Observatory (1973) is joined by a new breed of enclosure for the robotic
2 metre Faulkes Telescope South, completed in 2004.
(*Kristin Fiegert, Anglo-Australian Observatory*)

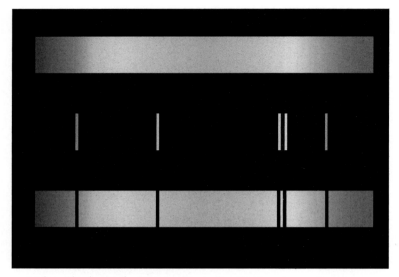

Light with a bar-code. Diagram illustrating the three types of spectrum:
- *Continuous spectrum* (top) from a hot object such as a lamp-bulb filament.
- *Emission-line spectrum* (middle) from a glowing gas, as in a flame. The pattern of bright lines identifies the gas.
- *Absorption-line spectrum* (bottom) from a hot object seen through a cooler gas. (In this case, it is the same gas as in the emission-line spectrum.)

Acting like an immense natural telescope, the rich galaxy cluster Abell 2218 forms distorted arc-like images of very remote galaxies by the process of gravitational lensing
(NASA, A. Fruchter and the ERO Team, (STScI, ST-ECF))

Locomotive-builder Andrew Barclay believed there was nothing wrong
with his telescope and that Saturn really did look like a
half-eaten apple.

In the tetchy correspondence that followed, Barclay
revealed that he had 'spent over £10,000 to find out how to
finish metallic speculums [sic] and mix and cast the metals.
I have made over 2000 experiments in connection therewith'.
Given that this comment was made at a time when Barclay
was in dire financial straits, it seems to betray not only that he
was obstinate, but also that he was stupid.

In 1800, all that lay in the distant future. The year following
Napoleon Bonaparte's *coup d'état* in France saw the political
map of Europe changing on an almost daily basis, and the
travails of astronomers were of little consequence in the wider
scheme of things. But the refracting telescope was also on the
verge of a revolution—one that would place British telescope-
makers at a grave disadvantage compared with their European
counterparts.

At issue was the availability of pieces of flint glass big
enough to make the concave elements of achromatic telescope

objectives with diameters larger than about 4 inches (10 cm). What little material could be bought was full of imperfections—bubbles and veins of non-uniformity—and it was expensive. As late as the 1820s, a Scottish astronomer called Thomas Dick noted that:

> Mr Tulley of Islington, who has long been distinguished as a maker of excellent achromatic instruments, showed me about six years ago a rude piece of flint glass about five inches [13 cm] in diameter, intended for the concave lens for an achromatic object glass, for which he had paid eight guineas.

That was an awful lot of money for an unworked lump of glass. Part of the problem was that the Government levied a tax on glass which, by 1825, had reached 98 shillings per hundredweight—a crippling excise. This 'window tax' was not finally abolished until 1851.

The bigger problem, however, was the inability of British glass manufacturers to produce blocks of flint glass in large sizes with anything like the required degree of homogeneity. Flint glass gets its useful refractive properties from the presence of lead—which has an unfortunate tendency to sink to the bottom of the molten glass and ruin its uniformity. So acutely was this problem felt that in 1824, at the Government's own behest, the Royal Society formed a committee to investigate ways of improving glass for optical purposes.

On the Continent, however, things were different. During the late 1790s, a Swiss cabinet-maker called Pierre Louis Guinand (1748–1824) had succeeded in producing reasonably acceptable flint glass lens blanks up to 5 inches (13 cm) in diameter, the result of years of painstaking experimentation above and beyond the call of cabinet-making. But in 1805, he perfected the technique of using a fireclay stirrer to prevent the lead component from sinking to the bottom of the melt. This breakthrough led to a succession of stunningly uniform flint glass blocks, setting the stage for a quantum leap in the technology of the refracting telescope.

In 1806, Guinand was persuaded to leave his native Switzerland and move to Benediktbeuern, south of Munich. On the site of the little town's old monastery was a glassworks, set up to provide raw material for a fledgling optical instrument factory in Munich. This Mathematical-Mechanical Institute, as it styled itself, had been established principally to manufacture surveying instruments with an eye to the military equipment market. The firm traded under the rather unmemorable names of its founders—Reichenbach, Utzschneider and Liebherr. Within a few years (and with a different name), it would become world famous, heralding the rise of the German precision instrument industry and the end of the near-monopoly that had been enjoyed by British firms like Dollond and Ramsden. It was not, however, Pierre Guinand who achieved that transformation.

WHIZZ KID

Guinand had been recruited by the senior partner in the organisation, a wealthy Munich lawyer and entrepreneur called Joseph von Utzschneider. Five years earlier, in July 1801, Utzschneider had had an extraordinary encounter with a fourteen-year-old orphan who had been pulled from the wreckage of a collapsed house in Munich. This unfortunate lad had lost both his parents a few years previously, and had subsequently been apprenticed in conditions of near-slavery to an ornamental glass-cutter and mirror-maker called

Philipp Weischelberger. It was Weischelberger's house that had collapsed, tragically killing his wife and trapping his young apprentice for several hours before he could be pulled from the rubble.

Utzschneider had been present at the dramatic rescue, and had been impressed with the boy's pluck and thirst for knowledge. Along with no less a figure than the Elector himself, Prince Maximilian Joseph, Utzschneider took an interest in the youngster's welfare, providing him with books on mathematics and physics. As it turned out, the boy had little choice but to continue working for his master, the miserly Weischelberger, but in his spare time he educated himself in the ways of optics and soon became thoroughly proficient. In fact, he demonstrated such brilliance in the field that in May 1806 he was rescued for a second time—headhunted to work at Utzschneider's Mathematical-Mechanical Institute.

Who was this talented young man? Joseph von Fraunhofer was born in Straubing in eastern Bavaria, on 6 March 1787. He was the youngest of eleven children—of whom only a pitiful four survived beyond infancy—and was himself not a particularly robust child. His father was a glass-cutter, and Joseph no doubt expected that one day he would follow the same trade, even after his father's premature death. But his arrival in Utzschneider's Institute at the tender age of 19 marked the beginning of a career that quickly became dazzlingly spectacular. He soon learned all aspects of the production of lenses, from the mathematical calculation of their surfaces to the final optical polishing—and then went on to become expert in mechanical design and, most significantly, glass manufacture.

It was from Pierre Guinand that Fraunhofer learned the tricks of optical glass-making. From the outset there had been a marked reluctance on Guinand's part to share his specialised knowledge, but in 1809 he was firmly instructed to do so by Herr Utzschneider—the boss himself. Fraunhofer, now a junior partner in the firm, recognised deficiencies in Guinand's production methods, and became increasingly critical. Not surprisingly, a deep animosity began to develop between the elderly glass-maker and the 22-year-old whizz kid. Matters came to a head in May 1814, when Guinand finally lost patience and returned to Switzerland. But by then, Fraunhofer had every aspect of optical technology—including glass-making—firmly in his grasp.

As if in anticipation of the brevity of his career, Fraunhofer now soared in his optical studies. His thinking encompassed novel designs for telescope objectives, new eyepiece designs, new ways of measuring small angles between celestial objects, and even researches into the very nature of light. His lenses benefited from new glass-production methods, and began to appear in steadily larger sizes. He had already produced a 7 inch (18 cm) telescope objective in 1812, and seven years later he completed his masterpiece—an exquisite 9.5 inch (24 cm) objective lens for the Russian observatory at Dorpat (now Tartu, Estonia).

The instrument that Fraunhofer built to accommodate this giant object-glass was for several years the largest refracting telescope in the world, but it was notable for more than just its size. Its 14 ft (4.3 m) long polished wooden tube was carried on a novel mounting that set new standards in efficiency and convenience. So effective were its basic principles that small telescopes are still mounted this way today.

Imagine two steel axles fixed at right angles to one another

Joseph Fraunhofer's 1824 masterpiece, the 9.5 inch (24 cm) refracting telescope for the Dorpat Observatory, which set the standard for the future.

in the form of a T. If the longer one is inclined so as to be parallel to the Earth's axis, it only needs to be turned at the same rate as the Earth to keep the telescope pointing in the same direction in space. (The Earth makes one revolution every 23 hours 56 minutes—a day reckoned by the stars rather than the Sun.) Because it points towards the celestial pole, this is called the 'polar axis', and it turns in heavy-duty bearings on a solid support.

The shorter axis—the top of the T—carries the telescope tube on one end and usually a counterweight at the other to balance everything up. An observer can move the telescope about this axis to find the right star, locking it at the correct setting once the object has been acquired. The star can then be tracked across the sky by rotating only the polar axis. Because it moves in the latitude-like sense that astronomers call 'declination', this second axis is known as the declination axis.

The principle of making one axis parallel to the Earth's had been adopted by Tycho Brahe nearly two and a half centuries earlier in his Great Equatorial Armillary (see Chapter 2), but in the elegant form developed by Fraunhofer it is universally known as a German equatorial mounting. And Fraunhofer added a further refinement—a weight-driven clock to keep the telescope exactly in alignment with the stars as the Earth turned under it. It was nothing short of brilliant.

Wilhelm Struve (1793–1864), director of the Dorpat Observatory and a famous double-star observer, almost wept when the great refractor was first turned on the sky on 16 November 1824. He wrote that he was:

> ... undetermined which to admire most, the beauty and elegance of the workmanship in its most minute parts, the propriety of its construction, the ingenious mechanism for moving it, or the incomparable optical power of the telescope, and the precision with which objects are defined.

The telescope served him well—he used it to measure more than 3000 double stars, many of which were separated by less than the elusive one arcsecond. Happily, the great instrument still exists, having been painstakingly restored in 1993.

It was with another Fraunhofer telescope that Friedrich Wilhelm Bessel (1784–1846) detected for the first time the minute back-and-forth shift in a star's position caused by the Earth's motion around the Sun. This instrument was made in the form of a heliometer with a divided objective for precise angular measurements. Bessel was director of the Königsberg Observatory when he measured this so-called 'parallax' (and hence the distance) of an inconspicuous star called 61 Cygni in 1838. It was the final clinching proof—if any was still needed—of the Copernican model of the Solar System, and of the vastness of interstellar space beyond.

By 1825, Joseph Fraunhofer was at the pinnacle of his career. He was a director of Utzschneider and Fraunhofer (the name under which the Munich Institute now traded). He was a corresponding member of the Munich Academy. He was a member of the Civil Order of the Bavarian Crown, and had been knighted for his services to optics in August 1824.

But he was not a well man. His boyhood frailty seemed to have caught up with him again—perhaps as a result of his constant exposure to the toxic fumes of the glass furnaces. Suffering from tuberculosis, Fraunhofer withdrew to his home during the winter of 1825–26, eventually carrying out his work from his sick-bed. On 7 June 1826, he died. He was only 39.

Fraunhofer's burial at the Südfriedhof (South Cemetery) in Munich was accompanied by a State funeral. His final resting place is next to his erstwhile colleague Georg Friedrich von Reichenbach, one of the three founders of the Munich Institute. A century and three-quarters later, on a chilly Munich afternoon, a handful of Fraunhofer's academic descendants took a few minutes out of their high-tech symposium on 'Power Telescopes and Instrumentation into the New Millennium', and paid homage to their great German forebear.

ALL-OUT WAR

When Pierre Guinand left the Munich Institute in 1814, he was admonished not to carry out further work in glass-making—in return for which he would receive a handsome pension. But Guinand was a born experimenter. When Utzschneider heard that he had resumed his work in glass manufacture, the pension was smartly terminated. From then on, Guinand felt no qualms about seeking new improvements in glass production—especially when he stumbled on a new technique of making large optical disks by pressing softened glass into a circular mould.

By this method, Guinand produced a number of very large glass blanks that ended up in the capable hands of Continental opticians. A few were bought by a talented Parisian lens-maker called Robert-Aglae Cauchoix (1776–1845), and it was one of these glasses that innocently sparked a no-holds-barred contest between two of the most prominent British astronomers of the day.

Unlike Guinand's disagreement with the youthful Joseph Fraunhofer, this controversy was very public—and both participants displayed exceptionally bad behaviour. The tragic results were that potentially the most significant refracting telescope in England was never used for astronomical research, and that gifted scientists wasted their time in pursuing the dispute to the exclusion of almost everything else.

The combatants in the engagement were really two groups of people, but the two individuals at its centre were so deeply entrenched in battle that it is easy to overlook the rest. Those gentlemen were, in the blue corner, Sir James South (1785–1866), and in the red, the Reverend Richard Sheepshanks (1794–1853).

South was a gifted amateur astronomer of independent means—a surgeon by profession who had married a wealthy heiress. His work on double-star measurement was so highly respected that he had been knighted in the 1820s to stop him taking his skills and his impressive suite of astronomical instruments to France. He had stayed in Britain, but had

Sir James South and Rev. Richard Sheepshanks.

become progressively more outspoken in his criticism of the establishment. His targets included the Royal Observatory (for deficiencies in the *Nautical Almanac*) and the Royal Society. He was, however, an enthusiastic founding member of the Astronomical Society (which in 1831 became the Royal Astronomical Society), and was its president from 1829 to 1831.

Richard Sheepshanks was a very different individual. He was the forthright son of a Yorkshire mill-owner with a Cambridge education in mathematics and a scornful attitude to those less talented than himself. His accomplishments didn't stop at mathematics; he studied law and was called to the Bar, and also took holy orders in the Church of England. He, too, was an enthusiastic supporter of the fledgling Astronomical Society, and was its secretary from 1829 to 1831.

It will not have escaped your attention that these two men held office in the Society at the same time, and that was no accident. The man who had proposed Sir James South as president—one Edward Stratford—felt that South needed a person as strong-willed as himself in the secretary's job to keep things in order, and proposed Sheepshanks. The two office-bearers took an instant dislike to one another. 'Sir James does not know a sine from a cosine, and is not able to use a table of logarithms for the simplest computation', Sheepshanks wrote later. South was similarly contemptuous of the young upstart—no matter how brilliant he might be.

Thus began a lifetime of acrimony between the two men. But what turned it into the all-out war it eventually became was Sir James' procurement, late in 1829, of an 11.75 inch (30 cm) diameter objective lens from the great Robert-Aglae Cauchoix. The lens was of exquisite quality, and the largest in the country at the time. South placed it in the hands of his then friend, the instrument-maker Edward Troughton (1756–1835), asking him to turn it into Britain's most powerful telescope. This Troughton and his business partner, William Simms (1793–1860), agreed to do, but the difficulties started as soon as the order had been placed.

South had requested that the telescope should be placed on a similar mounting to his existing 3.75 inch (9.5 cm) telescope, from which he had had excellent service. Troughton, however, believed that simply scaling up the design would not produce the necessary stability, and insisted on a new arrangement altogether. Reluctantly, South agreed, but continually interfered with the work as it progressed. The design adopted was of a pattern known today as an English equatorial mounting. It is an older form than Fraunhofer's German mounting, and differs from it in having a much longer polar axis, which is supported on bearings at both ends. The declination axis takes the form of a crosspiece about halfway between the two bearings. In South's instrument, the polar axis was made of wood, as was common at the time.

When the completed telescope was installed at South's residence in Kensington, London, in 1831, there was an accident that resulted in the body of the instrument being dropped on to the ground, damaging both the telescope and the framework of the dome. Fortunately, the precious 30 cm lens was not in the instrument at the time.

Once it was installed, however, the telescope was found to wander in the sky, with stars floating unsatisfactorily across the field of view. This was a fatal flaw in an instrument intended for the accurate measurement of double-star positions. Acrimony followed as Troughton and Simms attempted to remedy the fault while South hindered their workmen on site. Troughton himself was rather an obstinate man, and refused to consider reverting to the old design. When, in May 1832, South complained in a formal letter that Troughton had constructed a 'useless pile', the die was cast. The elderly Troughton was urged by a friend to take legal action to recover the payment owed him by South, and was assured by this same friend that he himself would provide legal advice. There are no prizes for guessing that Troughton's helpful adviser was a certain Richard Sheepshanks.

Fortunately for historians, there are copious records of the salvos that went backwards and forwards between Troughton, Sheepshanks, et al. on the one hand (who also brought on board the Astronomer Royal, Sir George Airy), and South on the other (whose allies eventually included Sir Charles Babbage). They take the form of letters, legal notes, even records kept in an exercise book by William Simms. They make depressing reading, demonstrating quite clearly that the real issue of the telescope was lost in a storm of personal vendettas, none of which was greater than that between Sheepshanks and South.

Edward Troughton died in 1835 with the issue still unresolved. Three years later, however, on 15 December 1838, there was a legal victory for his firm when Troughton and Simms were awarded the costs due to them by South, a sum of £1470. Predictably, South was enraged—not so much by his financial loss, but because he saw the hand of Sheepshanks in all his troubles. He resolved to humiliate his adversaries in any way he could and, in 1839, destroyed the main parts of the telescope mounting. He then put up posters throughout the district inviting:

> Mahogany Door Knob, Drawer Knob and Ball Turners;
> Stool, Button, Lucifer Match, Snuff Box Makers,
> and Dealers in Fire Wood and old Iron . . .

to an auction of:

> A quantity of Mahogany, other Wood and Iron, being the Polar
> Axis of the
> GREAT EQUATORIAL INSTRUMENT
> made for the Kensington Observatory by Messrs.
> TROUGHTON AND SIMMS.

There was to be no denying him his vengeance.

South preserved the brasswork of the instrument, but almost exactly four years after the legal settlement, held another sale with a similar invitation to the humblest categories of tradespeople in the district. This time, he went further. The advertisement was addressed to:

> Shy-cock Toy Makers, Smoke Jack Makers,
> Mock Coin Makers, Dealers in Old Metals . . .

and invited them to an auction of the metal parts of the:

GREAT EQUATORIAL INSTRUMENT
made for the Kensington Observatory by Messrs.
TROUGHTON AND SIMMS,
the Wooden Polar Axis of which, by the same Artists, and its
Botchings cobbled up by their Assistants,
MR. AIRY and the REV'D R. SHEEPSHANKS
were, in consequence of public advertisement on the 8th of July
1839, purchased by divers Venders of Old Clothes, and Licenced
Dealers in Dead Cows and Horses, &c. &c . . .

The melancholy scene at Sir James South's observatory on 8 July 1839,
when he auctioned the broken parts of his 'useless Twenty feet
equatorial'.

Early twentieth-century historians cited these outrageous posters as proof that South was mentally deranged, and very much the villain of the piece. But more recently, the historian Michael Hoskin has argued that the picture was far from black and white, and that blame for the episode should rightly be divided between South and Sheepshanks.

And it is true that the battle did not stop with the auction sales, but raged on until Sheepshanks' death in 1853. Even then, South could not resist a final shot when he published a letter aimed at those members of the Royal Astronomical Society who had written a warm obituary for his adversary.

And what of the marvellous Cauchoix lens? It languished at Sir James South's home—unused except for occasional celestial sightseeing with a temporary wooden stand—until he gave it to Dublin University in 1863. But by then, as Hoskin has put it, its moment had passed. What a tragedy for astronomy.

12

LEVIATHANS

MONSTERS WITH METAL MIRRORS

Sir James South's choice of Dublin University to receive his precious Cauchoix lens in 1863 was no idle whim. The middle decades of the nineteenth century had seen Ireland emerge as a major force in astronomy, thanks to a group of innovative pioneers who had seized any and every available opportunity. It was no accident, therefore, that by then the world's largest telescope lay within Ireland's green pastures.

The glue that held together this enterprising network of astronomers came in the form of Thomas Romney Robinson (1792–1882), a gifted physicist and astronomer who additionally boasted a doctorate in divinity among his qualifications. From 1823 until the end of his long life, Robinson was Director of the Armagh Observatory in what is now Northern Ireland. Among his many friends was the engineer of the group, a man by the name of Thomas Grubb (1800–1878). This Quaker descendent of Cromwellian settlers had been born in Waterford, and it seems likely that his own early interest in astronomy and optical instruments had

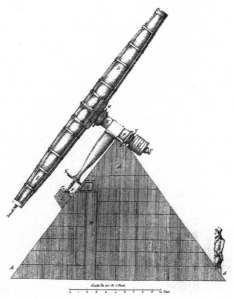

The imposing Markree telescope, built in 1834 by Thomas Grubb to accommodate Edward Cooper's 13.3 inch (34 cm) lens.

stimulated his friendship with Robinson.

By 1832, Grubb had a small engineering business in Dublin, producing goods ranging from machine tools to cast-iron billiard tables—and, occasionally, small telescopes. While billiards might have put bread on the table, it was the telescopes that fired Grubb's imagination and challenged his engineering skills. He soon had the opportunity to put those skills to the test for, in that same year, a wealthy County Sligo landowner and distinguished amateur scientist called Edward Cooper (1798–1863) ordered a telescope to house a 13.3 inch (34 cm) objective lens he had recently bought from the redoubtable Robert-Aglae Cauchoix. With a diameter an inch and a half greater than South's lens by the same maker, this objective was the largest in existence at the time.

Grubb's mounting for it was a triumph. Erected at Markree, Cooper's estate, in April 1834, the telescope's 25 ft (7.6 m) long tube was pointed by means of a sturdy equatorial built on Fraunhofer's German pattern, complete with clock drive (see Chapter 11). It had to withstand the elements, for there was no protective dome, but the whole structure was supported on an unusual triangular pedestal made of black marble. This spectacular engineering eventually yielded equally spectacular results in the shape of a catalogue of positions for some 660 000 stars. After Cooper's death, the telescope had a rather chequered career. It eventually finished up at the Manila Observatory in the Philippines, where the venerable Cauchoix lens is still used today as part of a solar telescope.

Hot on the heels of Grubb's success with the Markree refractor came an order for another instrument—this time a 15 inch (38 cm) reflecting telescope—from his friend Romney Robinson at Armagh. Completed in 1835, this instrument has a number of claims to fame. First and foremost, it was the earliest large reflector to be mounted equatorially, the clock-driven German-type mounting being essentially a copy of the Markree telescope's. (It may even have predated the Markree instrument, possibly having been built as a prototype.) But of no less significance—particularly in the light of what was to follow—was the ingenious device Grubb designed to hold the telescope's speculum-metal mirror.

Until that time, mirrors for reflecting telescopes had simply been placed in a box at the bottom of the tube with little thought given to the way this rough and ready support might affect the perfection of the reflecting surface. With the Armagh telescope, Grubb introduced the idea of a steel 'mirror cell' (a circular box) containing a system of supporting levers to allow the mirror to float in equilibrium. It was a very successful scheme, eliminating excess pressure on any part of the mirror's rear surface, and its computer-controlled descendants carry the mirrors of all the world's great reflectors today. Happily, the pioneering Armagh telescope with its innovative mirror cell still exists, and has recently been restored in the workshop of Thomas Grubb's modern-day counterpart, David Sinden of Newcastle-upon-Tyne.

The success of this and other telescopes built by Grubb enhanced his reputation to such an extent that by 1840 he had become Engineer to the Bank of Ireland. Though engraving banknotes might appear a far cry from building telescopes, similar precision techniques were required in both fields and it seems that from then on the versatile Grubb managed to combine them without difficulty.

Grubb's innovative reflector of 1835 for the Armagh Observatory, restored in 2003 by David Sinden.

It must have been some time earlier that Thomas Grubb had been introduced to the most famous member of the Irish astronomical fraternity—no doubt through the good offices of Romney Robinson. William Parsons (1800–1867), the hereditary Lord Oxmantown, was a man of great scientific ability who possessed all the resources necessary to fulfil his not-inconsiderable ambitions.

In 1841, he became the third Earl of Rosse, inheriting the rich ancestral estate of Birr Castle at Parsonstown (the town of Birr today) in the very heart of Ireland. His fortune had already been assured in 1836 by his marriage to Mary Field, a wealthy heiress from the village of Heaton, now a suburb of Bradford in Yorkshire. Heaton still celebrates the union, boasting a Rossefield Road and a Parsons Road. Mary was herself a notable figure—an early pioneer of photography who became the inaugural recipient of the Photographic Society of Ireland's Silver Medal in 1859.

The future Lord Rosse had begun his experiments in telescope-making back in 1827. From the outset he had endeavoured to keep the scientific world informed of his progress—unlike some of his predecessors—and had published his first results on grinding and polishing metal mirrors in 1828. He pioneered a novel technique of building up his mirrors from hollow sectors, like slices of a cake, in order to avoid the weight of solid speculum metal. In this way, he made successive lightweight mirrors for 15 inch (38 cm), 24 inch (61 cm) and 36 inch (91 cm) aperture telescopes. He also built a steam-driven polishing machine to surface them.

The last of these mirrors was completed in 1839, and Rosse fitted it into a Newtonian telescope on a large altazimuth mounting similar to Herschel's Forty-foot (see Chapter 10). Significantly, he adopted Thomas Grubb's lever-support system for his mirrors, although his acknowledgment of this in the Royal Society's *Philosophical Transactions* was little more than a passing reference:

> ... in supporting specula of three feet diameter I have availed myself of the suggestion of a clever Dublin artist, Mr Grubb, and, at the expense of a little more complication, have substituted nine plates for the three, resting on points supported by levers ...

The segmented 36 inch mirror worked remarkably well, and Rosse carried out tests to establish its capabilities for its intended task—resolving nebulae into individual stars. If that sounds familiar, your memory is doing a fine job, for it is exactly what Herschel had been up to more than half a century earlier. The fact was that despite the heroic efforts of the Herschel family, the riddle of the composition of nebulae had not been solved. It was Lord Rosse's intention to decipher it once and for all.

With that in mind, Rosse determined the following year to cast a solid 36 inch mirror to compare with the segmented one. Once again, he was extraordinarily successful, and along with two distinguished visiting astronomers was able to compare the two mirrors side by side late in 1840. The tests established the excellence of both, but demonstrated quite clearly that the solid one had the edge—whereupon Rosse promptly lost interest in lightweight segmented mirrors.

The tests of November 1840 are interesting for more than just the results, however. Detailed accounts reveal that the observers had a constant struggle with the weather, their work frequently being marred by poor seeing and curtailed by cloud. Unfortunately, such conditions eventually turned out to be typical of the low-lying Birr district and the nearby Bog of Allen.

Lord Rosse's eminent visitors were noteworthy, too. It would probably come as no surprise that one of them was Romney Robinson, but his colleague (and close friend) was none other than the infamous Sir James South. The historian Michael Hoskin has noted with justifiable amazement that during this period (little more than a year after South's first scandalous auction and two years before his second), Sir James was a model of good behaviour, providing wise counsel to both Lord Rosse and the Reverend Doctor Robinson. Such was his disposition when the Irish Sea separated him from his arch-foe, Richard Sheepshanks.

Robinson and South were together at Birr Castle again in 1845, when Lord Rosse turned his next telescope on the sky. But this was no 3 ft diameter drainpipe, suspended in a spindly tower that looked more like a colliery pit-head. Rosse's next telescope was a 6 ft (1.8 m) monster, the largest optical instrument that had ever existed—and it pointed its gaping mouth at the stars from within great walls of solid masonry.

SPIRAL STRUCTURE

The 'Leviathan of Parsonstown' had its origin in the successful tests of the 3 ft (91 cm) mirror. Fired with ambition, Lord Rosse immediately resolved to build an instrument of twice the aperture—and began work forthwith. His first attempt to cast a 6 ft (1.8 m) metal mirror took place on the evening of 13 April 1842 in the open air outside Birr Castle. Romney Robinson, present for the spectacle, was moved almost to poetry:

> The sublime beauty can never by forgotten by those who were so fortunate as to be present. Above, the sky, crowded with stars and

illuminated by a most brilliant moon, seemed to look down auspiciously on their work. Below, the furnaces poured out huge columns of nearly monochromatic yellow flame, and the ignited crucibles [of molten metal] during their passage through the air were fountains of red light, producing on the towers of the castle and the foliage of the trees, such accidents of colour and shade as might almost transport fancy to the planets of a contrasted double star.

Robinson, more than most, was conscious of the enormous significance of what was taking place before his eyes. To him, this dramatic scene must have encapsulated all the technological and scientific aspirations of the era—perhaps in the same way as the launch of a spacecraft might today.

Lord Rosse took every precaution in the casting process—including the provision of a large annealing oven, built into the castle moat. This allowed the 4 tonne mirror to be cooled slowly and in a controlled manner, avoiding the risk of internal stresses that might lead to cracks. It took no less than sixteen weeks to cool. In the event, the mirror did crack—but it was not due to faulty cooling, rather the result of an accident when the reflective front surface was being ground.

Undaunted, the Earl cast a second mirror, then a third . . . and eventually, a fifth. Of these, the second and fifth successfully underwent the steam-driven grinding and polishing process—and both took their turns in the telescope's gigantic mirror cell with its bed of supporting levers when the instrument was finally completed.

Lord Rosse adopted a very different structure from that of the 3 ft telescope to support the 56 ft (17.1 m) long wooden tube of his Leviathan. It hung between two massive walls 72 ft (21.9 m) long and 56 ft (17.1 m) high to protect it from the

Lord Rosse's monumental 6 ft (1.8 m) aperture reflecting telescope, the Leviathan of Parsonstown.

wind. The walls were aligned north–south, and sufficiently far apart (24 ft, or 7.3 m) that the telescope could track east-to-west on an object for an hour or so as it crossed the meridian—the imaginary line in the sky passing directly overhead between the north and south points of the compass.

To cope with the varying height of stars above the horizon (their altitude), the telescope could be raised on its supporting chains to the vertical position—or tilted on its back as far as the north pole of the sky. The mounting was certainly not an equatorial, but with the help of two assistants, the massive instrument could be made to track the stars reasonably well. Once again, Lord Rosse adopted the Newtonian layout, and the eyepiece was accessed from movable observing galleries supported high on the western wall. A convenient eyepiece-slide allowed an observer to change from low to high magnification instantaneously.

By February 1845, the Leviathan was ready for use, and Rosse, Robinson and South girded their loins for a fresh assault on the nebulae. As it turned out, it was more than just their loins that needed girding, for the winter weather was totally uncooperative, allowing them only enough starlight to establish that the mirror (the second casting) was of very high quality. Rosse must have been bitterly disappointed: his telescope had cost some £12 000 to build, and he was anxious to see a scientific return on his investment.

It was not until early March that the weather improved and systematic observing could begin. By the middle of that month, the intrepid celestial explorers had seen enough to convince themselves that all nebulae could indeed be resolved into individual stars—and were duly congratulating one another. That they were quite wrong was not finally proved until the work of William Huggins in the 1860s (see Chapter 14).

Something new did emerge, however: a subtle detail that no other telescope in the world had been large enough to reveal. It very quickly overtook the resolvability of nebulae as Lord Rosse's driving passion, and spawned another mystery that had to wait until the 1920s to be unravelled. One of the nebulae—an object with the unromantic name of M51 in the constellation Canes Venatici—had a curious spiral structure. It had the unmistakable appearance of a whirlpool, fascinating the observers and 'indicating modes of dynamic action never before contemplated in celestial mechanics'—as Robinson put it. They were right to be astonished. They had discovered the first known example of a spiral galaxy.

The Leviathan of Parsonstown was eventually used to find more than sixty of these 'spiral nebulae', as they were called, and that remains its greatest achievement. But the vast majority of them were found when the telescope was more than a decade old, its mirrors no longer in their prime. An unforeseen catastrophe had brought all normal life in Ireland to a standstill, interrupting the Leviathan's work along with everything else. During the potato famine of 1845–48, almost a million people starved and another million were forced to emigrate. Lord Rosse, as a responsible landowner, had to devote all his attention to his estate—as did Mary, who earned the particular affection of the local population by her efforts to alleviate their suffering. But the Government in London, responsible for Ireland since the *Act of Union* of 1801, was nowhere near as fastidious. With Anglo-Irish relations already strained, the seeds of revolution were sown amid the blighted remnants of the failed potato crop.

It was not until after the death of Lord Rosse's eldest son, Laurence (the fourth Earl), in 1908 that the great telescope began to show serious signs of decay. Soon it was dismantled, leaving little more than the derelict tube and masonry walls. When Henry King completed his monumental history of telescopes in 1955, that was the end of the Leviathan's story. Today, however, there is a remarkable postscript. Between 1996 and 1998, the great instrument was sympathetically restored to full working order, with a new aluminium mirror and modern hydraulic mechanisms to move it. Thanks to the vision and energy of William Brendan Parsons, seventh Earl of Rosse, the telescope now represents an exciting fusion of old and new technology, a shining example of the best of both worlds.

COMFORT AND JOY

Notwithstanding Ireland's leading position in the mid-nineteenth-century super-telescope league, intriguing developments were afoot on the other side of the Irish Sea. At Patricroft, near Manchester, a Scottish engineer named James Nasmyth (1808–1890) was investing much of his spare time in the construction of reflecting telescopes with speculum-metal mirrors. Like Andrew Barclay, his reputation had been made in heavy engineering rather than astronomy—Nasmyth is best known for the invention of the steam hammer in 1839. Unlike the hapless Barclay, however, his telescopes really were something to write home about.

Nasmyth was an extraordinarily gifted engineer with a philanthropic bent, and several of his most important inventions were left unprotected by patents so that they might be widely adopted, improving the safety of workers. He was a noted raconteur, frequently telling the story of a bargee on the Bridgewater canal who had glimpsed him in his nightshirt

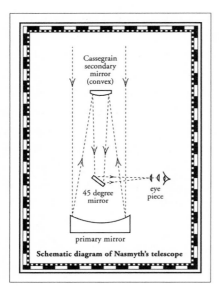

Cross-section of Nasmyth's telescope showing its Cassegrain form with an additional flat mirror to bring the light to the fixed eyepiece.

lifting a telescope out into his garden. The terrified boatman was certain he had seen a ghost carrying a coffin in its arms. No doubt it was Nasmyth's reckless love of life that also led to his enduring extramarital affair with a certain Flossie Russell, who bore him a daughter in 1859.

Nasmyth was an accomplished artist and pioneer photographer. His name lives on in astronomy, however, through the unique design of his largest telescope, a 20 inch (51 cm) reflector of advanced form completed in 1842. This instrument had a sheet-iron tube mounted in trunnions (bearings) that allowed it to swing vertically between two triangular pillars on a rotating turntable, an altazimuth arrangement similar to Herschel's Large Twenty-foot and Forty-foot telescopes. But in contrast with those archaic wooden structures, the engineering of Nasmyth's 20 inch had more in common with the mounting of a modern naval gun. The observer's seat, for example, was fixed to the turntable, rotating with it. And all the movements of the instrument could be controlled from conveniently placed handwheels.

Even more convenient was the fact that the observer always looked in the same direction, no matter where the telescope was pointing in the sky. This arrangement for making 'Gigantic Telescopes at once Easy and *comfortable*' (as Nasmyth described it) was achieved by means of a hybrid three-mirror optical system. In essence, it was the concave primary–convex secondary combination described by Cassegrain, but with the addition of a 45-degree flat mirror to direct the light beam out through the side of the tube rather than letting it go through a hole in the main mirror. The 45-degree mirror was aligned with the hollow elevation trunnion of the mounting, and this was where the eyepiece was placed—right in front of the observer's seat. It was a very neat arrangement, though at the cost of an extra, lacklustre speculum-metal surface.

Nasmyth used this telescope primarily to study the Moon, and eventually collaborated with one James Carpenter to write a book on the formation of the lunar craters (published in 1871). But it is for his elegant mounting that he is remembered, since most of today's largest telescopes use the same basic structure, with a 'Nasmyth focus' where heavy auxiliary instruments can be mounted. Sadly, in this electronic age, they no longer include a comfortable seat for the observer. The original telescope is preserved in London's Science Museum.

Among James Nasmyth's friends was another English 'grand amateur' astronomer who made a significant contribution to the development of reflecting telescopes. William Lassell (1799–1880) had made his money as a brewer—like the

James Nasmyth at the controls of the 'comfortable' 20 inch (51 cm) reflecting telescope he completed in 1842.

doughty Johannes Hevelius two centuries before—and by the 1840s he was an accomplished and experienced observer, specialising in the satellites of the outer planets (of which he discovered several).

Lassell eschewed the complexity of the Nasmyth layout, preferring the more straightforward Newtonian design, but he and Nasmyth collaborated on steam-powered polishing machines for telescope mirrors. Based in Liverpool, Lassell was within easy reach of Nasmyth by virtue of the new public railway that had linked Manchester with the Mersey seaport since 1830. It was not that much harder to get from Liverpool to Ireland and, in 1844, Lassell visited Parsonstown to take stock of Lord Rosse's progress with the 6 ft reflector. He had in mind a 24 inch (61 cm) telescope of his own, and was interested in all aspects of speculum fabrication and polishing.

Lassell's 24 inch telescope first saw starlight a couple of years later, and quickly reaped a major success. Less than two weeks after news of the discovery of the planet Neptune had reached London on 30 September 1846, Lassell found the planet's one large moon, Triton. It was an auspicious start; some six years later, the telescope's capabilities were further enhanced when he took it to Valetta in Malta, seeking clear skies and stable air.

Once back in murky Liverpool, Lassell followed the path trodden by Lord Rosse, resolving to double the aperture of this, his best telescope. The resulting 48 inch (1.2 m) instrument represents a true milestone in the history of the reflecting telescope. Perhaps its only major weakness was the continuing battle against tarnishing that always accompanied speculum-metal mirrors. But, following the practice established by Herschel nearly a century before, Lassell provided the instrument with two interchangeable 48 inch diameter mirrors, each weighing well over a tonne.

The innovation embodied in Lassell's telescope centred on its 37 ft (11.3 m) long tube, the support arrangements for the primary mirror, and the instrument's equatorial mounting.

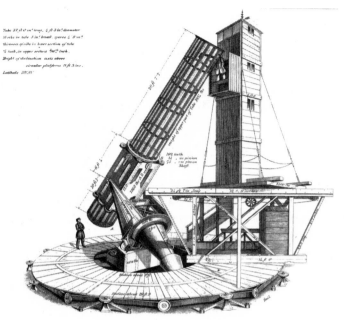

Lassell's elegant 48 inch (1.2 metre) aperture telescope of 1860 had many refinements, but its Newtonian layout necessitated a 'colossal sentry-box' for the observer.

Following a suggestion by Nasmyth that free circulation of air around the mirror might be important for temperature stability, Lassell made his tube as an open framework of iron slats—a thoroughly novel approach. Then, seeking further improvements over Grubb's mirror support system, he counterpoised his own mirror with a new system of balancing weights, today called an 'astatic' support.

Finally, realising that Fraunhofer's elegant German equatorial mounting was not terribly suitable for large, fat reflecting telescopes, he built what is now known as a fork mounting. Essentially, it was similar to Nasmyth's 20 inch mounting, but with the turntable replaced by a massive polar axis inclined at the angle of latitude of the observing site. Thus, the telescope could swing in declination (angular distance from the celestial equator) between the two prongs of a gigantic fork. Once again, Lassell chose the Newtonian arrangement for the optics.

Lassell finished this marvellous instrument in 1860 and, as before, moved it to Malta the next year. It performed beautifully, Lassell observing the heavens from the 'colossal sentry-box' that gave precarious access to the eyepiece at the top of the tube. His description of the Orion Nebula tallies well with modern images of this brightly shining cloud of gas and dust:

> Examined under good circumstances, with a [magnifying] power of 1018, the brightest parts of the nebula look like masses of wool . . . one layer seemingly laying partly over another, so as to give the idea of great thickness or depth in the stratum.

Lassell stayed for three years in Malta, observing planets and satellites and cataloguing some 600 nebulae. When he finally returned, it was to observe only with the 24 inch, for he never re-erected the larger telescope. He had other plans for it—ambitious plans that were, in the event, thwarted by the deliberations of a well-meaning but misguided committee.

13

HEARTBREAKER

THE GREAT SOUTHERN TELESCOPE

The metal-mirrored telescope reached its bittersweet apotheosis in a project that involved almost everyone we met in the last chapter. It carried the reflecting telescope tantalisingly close to its twentieth-century form, but brought in its wake controversy and recrimination—and eventually, in an extraordinary twist, high drama in our own time.

It was the Victorian scientific world's continuing obsession with the nature and structure of nebulae that spawned this ambitious project. Astronomers were painfully aware that the only systematic study of nebulae in the southern hemisphere had been carried out between 1834 and 1838 by John Herschel at the Cape of Good Hope, using his father's Large Twenty-foot reflector. This instrument, with its mirror diameter of 18.7 inches (47 cm), was the biggest that had been used in the south. But following the spectacular results on spiral nebulae that had come from Lord Rosse's mammoth 6 ft reflector in the 1840s, no one was in any doubt about the potential value of a new large telescope south of the equator.

Except, that is, most of the professional astronomers of the day. In April 1850, Her Majesty's Astronomer at the Cape, one Thomas Maclear (1794–1879), responded to an enquiry from Romney Robinson about the possibility of siting a large telescope there. His tone was almost indignant:

> Hitherto the examination of Nebulae with powerful instruments, also the measurement of double stars, have been left exclusively or nearly so to the zeal and resources of private astronomers. The public Observatories being devoted to the more immediate utilitarian branches of cataloguing and of improving the planetary theory by systematic observations . . .
>
> I feel that the standard work [systematic positional measurements with small telescopes] has prior claim and owing to accidental circumstances it is in a backward state in this hemisphere . . . the all-absorbing interest attending the mounting and working of a large Reflector would further retard and depress that which I am anxious as a duty to accomplish.

Notwithstanding Maclear's lack of enthusiasm, Robinson collaborated with Lord Rosse (then President of the Royal Society) to petition the Prime Minister to support a large southern telescope. This proposal came to nothing because the PM (Lord John Russell) never saw it. It was intercepted and sabotaged—possibly unintentionally—by another professional, George Airy (1801–1892), the Astronomer Royal.

A couple of years later, in 1852, a further attempt was initiated. This time, the project gained the support of the Royal Society, which followed its usual practice of forming a panel of experts to manage it. Numbered among the members of this Southern Telescope Committee were the movers and shakers of the large telescope world: Romney Robinson, Edward Cooper, Lord Rosse, James Nasmyth, William Lassell, John Herschel—and, now, an enthusiastic George Airy too.

The multi-talented Engineer to the Bank of Ireland, Thomas Grubb, was engaged as a consultant, and he proposed a telescope of 48 inch (1.2 m) aperture. Its radical design abandoned the well-tried Newtonian layout in favour of the Cassegrain form, in which the eyepiece is located behind the perforated primary mirror. This avoided having observers teetering precariously on high access platforms, and the arrangement of the telescope's English equatorial mounting (see Chapter 11) further ensured that the eyepiece position would never be far from the ground. Grubb estimated it could be built for some £5000, the price including two 48 inch speculum-metal mirrors, a polishing machine to maintain them, and a one-horsepower steam engine to drive it.

Buoyed by this ambitious proposal, the Southern Telescope Committee resolved on 5 July 1853 'that application be made to H.M. Government for the necessary funds'. Its enthusiastic members had every hope of success. But they could hardly have guessed that only days before, an event in distant lands had doomed the project to certain failure. Russia's occupation of Moldavia and Wallachia—part of modern-day Rumania—led the following year to the Crimean War of 1854–56. Under these circumstances, with Britain deeply involved in the conflict, there was no possibility of Government funding for the proposed new telescope.

That might have been the end of the story had it not been for something as unexpected in its own way as the outbreak of war. In 1851, significant reserves of gold had been discovered in the Australian colony of Victoria. As a result, the colony's population increased sevenfold within a decade, and its wealth flourished proportionately. Education in science and the arts prospered and, in 1855, the foundation Chair of Mathematics at Melbourne University was taken up by one William Parkinson Wilson (1826–1874), a former professor at Queen's College, Belfast.

It will probably come as no surprise that Wilson had worked in Ireland with the energetic Romney Robinson, and was an enthusiastic advocate of astronomy. What is more remarkable is that by 1860, the science had a staunch ally in the shape of the highest fiscal authority in the land—the Treasurer of Victoria. This man was George Verdon (1834–1896), an amateur astronomer who was *very* favourably disposed to the foundation of a government-funded observatory in Melbourne—which duly took place in June 1863.

Professor Wilson had already made overtures to the Mother Country about the possibility of the fabled southern reflector being installed in Victoria, and once the new Melbourne Observatory was up and running, a formal approach was made to the Royal Society. They responded enthusiastically to the idea, but declined to offer any funds for the project because

> ... the estimate which is entertained here of the enormous and increasing wealth of your colony would stand very much in the way of any proposition for supplementing the Colonial appropriation.

Nevertheless, the Royal Society reconvened the Southern Telescope Committee in the light of this spark of interest from a rich southern cousin. And, in its turn, the Government of Victoria responded generously, eventually allocating the full amount of £5000 to bring to the remote colony the world's most advanced telescope. The stage was set for untold new discoveries in the southern skies. Or so they all thought.

ENGINEERING MASTERPIECE

Once the funding had been established, events moved quickly. The reconvened Southern Telescope Committee recommended that Grubb's design be adopted essentially unchanged. That decision was controversial almost from the word go, for the mid-1850s had seen the emergence of a new technology for making telescope mirrors by coating a thin layer of silver on to a polished glass surface (see Chapter 15). This technique offered a number of advantages over conventional speculum-metal mirrors, including lightness (thereby permitting a much lighter structure for the telescope), vastly improved reflectivity, and the ability to recoat the mirror when it became tarnished rather than having to submit it to the exacting task of further optical polishing.

Despite these clear advantages (particularly the last, given that the telescope was destined to operate far from the experts who had built it), the committee elected to take the conservative approach and stick with speculum metal. They argued that silver-on-glass was an untried technique in a mirror of such large size, even though a 31 inch (80 cm) silvered-mirror telescope had been built for the Paris Observatory in 1862. It was a fateful decision, and without doubt was one of the contributors to the telescope's eventual failure to live up to expectations.

Further controversy came when William Lassell tried to persuade the committee to adopt the design of his own 48 inch (1.2 m) telescope—and later offered to *give* them the complete instrument. This was refused on the grounds of risks to the safety of observers using the lofty Newtonian eyepiece position, and the fact that Lassell's telescope needed considerable effort to steer it around the sky. That, at least, was a valid criticism, since Grubb's design could be moved with only one-thirtieth of the force that Lassell's required. There is no doubt that in the details of its mounting, Grubb's proposal was a masterpiece.

In February 1866 Thomas Grubb at last received the order to build the telescope. Heavily committed with his work for the Bank, he entrusted the Melbourne contract to his son Howard, then 21 years old. Three decades later, when Sir Howard Grubb headed one of the greatest instrument-building firms in the world, he recalled:

> . . . this Melbourne Telescope practically brought our optical works into being; for the moment the order was given, my father bought a piece of land at Rathmines [a suburb of Dublin] and erected temporary workshops, machinery and furnaces, suitable for casting the 4ft. speculum mirror.

Casting one of the 48 inch (1.2 metre) metal mirrors of the Great Melbourne Telescope.

By July, the fledgling optical works was ready for the mirrors to be cast. Grubb followed a similar process to that pioneered twenty-four years earlier by Lord Rosse, except that the work was done under cover to the accompaniment of a throbbing steam engine pumping air into the furnace. Such was the intense heat that the workers had to be clad in ungainly protective clothing, making them look to our eyes disconcertingly like members of the Ku Klux Klan.

The first attempt at casting, on 3 July, did not go smoothly. The furnace had been lit during the previous afternoon, and young Howard, having retired early, was rudely awakened:

> At 12.30 [am], a messenger rushed into the house with the cheerful news that the works were in flames; the almost red-hot chimney had set fire to the roof. I rose quicker than usual, and was presently playing on the blazing timbers with a garden hose. This was no good, so I just sawed away the beams from around the shaft, and then let the roof flare away.

That emergency having been dealt with, Grubb and his men toiled throughout the day to keep the furnace operating and, by 11 p.m., the two tonnes of molten metal were ready to be poured. The process took a mere six seconds, after which the white-hot mirror was transferred to the annealing oven. It is easy to imagine the relief they all felt at the end of it: 'At 1am on the 4th July I got home [recalled Grubb], having laboured continuously in that frightful place for twenty-four hours'. Unfortunately, their labours had been in vain, for the first casting turned out to be seriously flawed. But they learned from the process, and two successful mirrors were produced later in the year.

In comparison with the drama of casting the mirrors, the optical polishing and the fabrication of the telescope structure were routine processes that went relatively smoothly. By 17 February 1868, the completed telescope was ready for inspection by a subcommittee of the Royal Society panel. Erected in Grubb's yard at Rathmines, it provided spectacular views of the sky, and the subcommittee declared the instrument 'perfectly fit for the work for which it was designed'.

Within three months it was being dismantled for shipment across the Irish Sea, and it eventually left the port of Liverpool in July 1868 aboard the *Empress of the Seas*. At last, the great instrument was bound for the southern hemisphere—in the company of a cargo of salt, roofing slates and beer. It arrived in Melbourne on 6 November.

The erection of the telescope in its new home was not without incident, for the difference in latitude between Melbourne and Dublin had been incorrectly allowed for. Moreover, the first of the telescope's two mirrors had arrived in less than perfect condition, with a deterioration of its surface already evident. Another problem that quickly became apparent was the susceptibility of the spidery 9 m (30 ft) long tube to wind-shake—the telescope having been designed to operate without a protective dome by individuals blissfully unaware of Melbourne's prevailing weather.

Nevertheless, the instrument entered service in August 1869, and the Government Astronomer, Robert Ellery (1827–1908), expressed confidence that all its teething troubles were over. But he was wrong.

DECLINE AND DISASTER

It soon emerged that the Great Melbourne Telescope produced images of stars that were shaped more like the ace of clubs than well-defined points of light—a fault that was, in the event, quickly rectified by loosening the mirror cell's death-grip on its precious contents. But in March 1870, the *Australian Journal* published an article about the new instrument that was highly critical:

> This telescope, by means of which the public were led to believe that immense astronomical discoveries would be made, having been fixed in the Melbourne Observatory at the Botanical Gardens, has to a considerable extent turned out a failure . . . Mr Le Sueur, to whom the custody of the telescope is confided, says that the *cassegraine* form of construction is by far the very worst that could have been adopted. Rather a pleasant conclusion, certainly, after the expenditure of about fourteen thousand pounds upon the instrument and its fittings! . . . Altogether, the telescope may be looked upon as a gigantic philosophical blunder.

There is an uncanny similarity between these views and those aired exactly 120 years later, when the Hubble Space Telescope was discovered to have a defective primary mirror: 'outrageous waste of public funds . . .' and so on. But whereas confidence in the Hubble was restored in December 1993 by the addition of a small corrective mirror, the Melbourne telescope never recovered. Albert Le Sueur was, in fact, Ellery's Assistant Astronomer, and he was clearly disgruntled by what he saw as a major design flaw. After the not-inconsiderable achievement of repolishing the first mirror on site in July 1870, he resigned and moved back to England, taking with him all his expertise in that exacting task.

Grubb and Robinson, alarmed at the reports reaching them from the Antipodes, distanced themselves from the problems, blaming the methods used in Australia. And, indeed, throughout the life of the telescope, there were always problems with the mirrors, either due to tarnishing or poor imaging. A letter of 1875, for example, describes the star cluster Omega Centauri as having 'a very unsatisfactory appearance, the stars seemingly all running into each another, suggesting the appearance of a sago pudding'. That problem, at least, was easily fixed by an adjustment to the alignment of the mirror in its tube. In many respects, the telescope was a perfectly serviceable instrument—it produced the best lunar photographs of its day, for example—but the overall attitude towards it was by now uniformly gloomy.

Most commentators on the Great Melbourne Telescope have blamed its poor track record of discovery on the fact that it was designed and operated by a succession of committees rather than enthusiastic and single-minded individuals—the likes of Herschel and Rosse. The use of outmoded speculum-metal mirrors, with the attendant need for frequent repolishing, has also been cited. Certainly, the instrument was the last great telescope to use such mirrors.

But in a recent re-evaluation, the distinguished and much-loved Australian astronomer Ben Gascoigne reached a different conclusion:

> The real trouble was that the telescope had been built for the explicit purpose of making hand-and-eye drawings—'depictions'—of the southern nebulae . . . Even as the telescope was being built photography was promising to revolutionise the subject, and the advent of fast [sensitive] gelatin [photographic] plates in the 1870s made its adoption inevitable. That spelt the end of the pencil sketches. There was nothing else the telescope could do, neither photography nor spectroscopy, and it fell into disuse. Even if the mirror had been of silver-on-glass it would have made no difference.

So, had Le Sueur been right? Although the Cassegrain configuration was not itself intrinsically flawed, it had been adopted in a particularly extreme form in the Melbourne telescope, amplifying the mirror's focal length to a staggering 50.6 m (166 ft). This rendered it unusable for nebular photography simply because it diluted the faint light too much. It had been doomed from the moment Thomas Grubb had sketched out its design.

After many valiant efforts by Ellery in repolishing the mirrors, the Great Melbourne Telescope went into a long, slow decline—along with the Government Observatory that accommodated it—and was eventually closed down in March 1944. Whatever the real reason for its failure, the scathing assessment made in 1904 by the American astronomer and telescope-maker, George Ritchey, was not without foundation:

> I consider the failure of the Melbourne reflector to have been one of the greatest calamities in the history of instrumental astronomy; for by destroying confidence in the usefulness of great reflecting telescopes, it has hindered the development of this type of instrument, so wonderfully efficient for photographic and spectroscopic work, for nearly a third of a century.

And it was true: in the aftermath of Melbourne's heartbreak, the large refracting telescope had swung boldly back into prominence for its last hurrah. As the nineteenth century progressed through its final decades, there was no lack of confidence in the ability of telescope-makers to manufacture big lenses that would work properly.

After the closure of its host observatory, the Great Melbourne Telescope was sold to the Commonwealth Solar Observatory at Mount Stromlo, on the outskirts of Canberra. Although the instrument changed hands at scrap value—£500—it then underwent a series of metamorphoses that culminated in 1959 in the installation of a new 50 inch (1.3 m) Pyrex glass mirror. In this guise it did useful work in the field of photo-electric photometry (the electronic measurement of star brightnesses) in the hands of Ben Gascoigne and others. At about the same time, the institution itself metamorphosed into the Mount Stromlo Observatory of the Australian National University.

By the mid-1970s, however, the bearings of the telescope were completely worn out. But this great survivor, now more than a century old, was reincarnated yet again in the early 1990s to perform a specific scientific task—that of hunting for failed stars that might make up the mysterious 'dark matter' known to permeate the Universe. Fitted with a state-of-the-art electronic camera, it was spectacularly successful in this role. Although little of the instrument was by now original— specifically, only the mirror cell and parts of the polar and declination axes—it had, at last, lived up to its builders' expectations.

As Gascoigne had predicted in 1996, the Great Melbourne Telescope saw out the twentieth century 'in much better shape than it saw out the nineteenth'. A new programme to discover small, Pluto-like planets on the fringes of the Solar System was started in 2000 and, by 2002, plans were well advanced for the installation of a new camera that would be used in an ambi-tious survey of the entire southern sky. There seemed no limit to the reborn telescope's capabilities.

But then disaster struck. Mount Stromlo had been known to be vulnerable to bushfires since February 1952, when flames had swept through nearby pine plantations and damaged the Observatory's workshops. But no one was pre-pared for the events of 18 January 2003. On that black day,

The remains of the Great Melbourne Telescope after the devastating fire
of 18 January 2003. Grubb's polar axis is still recognisable.

a colossal firestorm in Canberra's south-western suburbs
claimed the lives of four people, and left more than 500 fami-
lies homeless. Once again, the tinder-dry pine forests around
Mount Stromlo burned furiously. This time they took with
them all the Observatory's heritage buildings, including the
domes of the Great Melbourne and five other historically
significant telescopes.

When I visited Mount Stromlo seven months later, the
remains were a pitiful sight. Their future still uncertain
while insurance claims were pending, all were surrounded by
safety barriers and bore signs warning of the danger of struc-
tural collapse. None was more heart-rending than the Great
Melbourne Telescope building itself. The aluminium dome
had melted in the intense heat, leaving the twisted wreckage
of the telescope now completely exposed to the elements.

Grubb's polar axis—still with its original maker's nameplate attached—lay rusting in the rains that heralded the end of Australia's once-in-a-hundred-years drought.

Today, there is little likelihood that the telescope will ever be restored to working order. A place in a museum seems the best that can be hoped for. How extraordinary that the end of this remarkable instrument should have come in unbidden flames, just as its beginning did. It seems never to have lost its capacity to break the hearts of its champions.

14

DREAM OPTICS

PERFECTING THE BIG REFRACTOR

On Good Friday 1868, seven weeks after the Royal Society's inspection of the newly finished Melbourne telescope, an event of great cultural significance took place in the north German city of Bremen. An up-and-coming young composer called Johannes Brahms directed the first performance of the monumental choral work that was to put his name firmly on the international musical map. *Ein deutsches Requiem* ('A German Requiem') was intended by the atheistic Brahms less as a conventional Latin Mass for the dead than a message of comfort for the bereaved—an aim that was entirely understandable, given that he himself had recently lost his mother. Moreover, its title reflected Brahms' fond use of the common tongue for the scriptural text rather than any particular nationalistic fervour.

This deeply thoughtful music, by turns sombre and radiantly uplifting, is built on foundations laid by Brahms' great predecessors: Joseph Haydn (1732–1809), Wolfgang Amadeus

Mozart (1756–1791), Ludwig van Beethoven (1770–1827) and Robert Schumann (1810–1856). Their musical signatures are clearly discernible in Brahms' work, intermingled with new perceptions to make an idiom uniquely his own. Like Newton, Johannes Brahms was standing on the shoulders of giants.

An orchestral score has much in common with a mathematical theory. Both are written in hieroglyphics that are meaningless to the uninitiated. Both take abstract ideas and develop them to a conclusion, shedding new light along the way and revealing fresh insights at the end. And, in their most inspired forms, both also demonstrate elegance and creativity in their smallest details as well as in their large-scale structure.

Brahms' music is like that. And so were the mathematical theories of a fellow countryman whose influence on German science was every bit as profound as Brahms' was on its culture. This man was Ernst Abbe, a mathematical physicist working at the University of Jena in Thuringia and specialising in optics. He and Brahms were near-contemporaries, the composer living from 1833 to 1897, the physicist from 1840 to 1905. It would be nice to think of them as friends—but it is unlikely they ever met.

Abbe made extraordinary contributions to theoretical optics and, just as Brahms had done, drew on the pioneering studies of his great predecessors: Carl Friedrich Gauss (1777–1855), Joseph von Fraunhofer (1787–1826), Joseph Petzval (1807–1891) and Ludwig von Seidel (1821–1896). In the hands of these men, the design of lenses for telescopes and other optical systems had progressed into a veritable symphony of mathematical ideas. However, just as a musical composition has to be turned into reality by performers, so must lenses be formed from real pieces of glass.

During the 1870s, Ernst Abbe realised that once again theory had far outstripped practice in the availability of optical glass.

> For years [he wrote later] we combined with sober optics a species of dream optics, in which combinations made of hypothetical glass, existing only in our imaginations, were employed to discuss the progress that might be achieved if the glass-makers could only be induced to adapt themselves to the advancing requirements of practical optics.

This time, it was not so much the availability of large chunks of glass that prevented the 'dream optics' from being turned into reality, but the existence of glasses with particular refractive characteristics. Such desirable properties would allow the manufacture of elegant lens designs with extraordinary capabilities. Field of view, colour correction, focal ratio (the ratio of focal length to diameter, which determines the level of image illumination), all could be optimised for particular applications.

Abbe was not a man to sit around dreaming for long. In January 1881, he met a young doctoral graduate in chemistry called Otto Schott (1851–1935), who had some new ideas concerning the manufacture of glass types with exotic properties. Abbe had already been working for a decade and a half with the gifted instrument-maker, Carl Zeiss (1816–1888), and in 1884, these three men—together with Zeiss' son, Roderich—founded the Schott glassworks in Jena. It was the beginning of modern glass technology, and it revolutionised the manufacture of optical instruments.

Now, not only could all kinds of lenses be made for both visual and photographic use, but optical prisms—blocks of glass with flat surfaces angled to transmit or reflect light—

could also be brought to a much higher degree of perfection. It was the smaller classes of instruments that reaped the benefits first. New types of microscopes and telescopes, and Abbe's amazingly successful prism binoculars, introduced in 1894, began to appear in ever-increasing numbers from the Zeiss factory. And then, during the first decade or so of the twentieth century, a clutch of superb astronomical refractors were built by Zeiss and other German firms, with apertures up to 80 cm (31.5 inches). They embodied all that was excellent in optical technology.

The paradox was, though, that the era of the large refracting telescope was by then virtually at an end. The biggest refractors in existence today had already been built in the USA by American manufacturers using lenses made of French glass. The credibility gap left by the failure of the Great Melbourne Telescope had closed rapidly. While there would still be a demand for lenses in instruments such as spectrographs and wide-angle star-cameras (astrographs), the silver-on-glass reflector had begun to assert its dominance as the large telescope of choice for astronomers.

SIFTING STARLIGHT

The large refracting telescope's gradual rise to perfection during the nineteenth century had begun with Fraunhofer's 9.5 inch (24 cm) Great Dorpat Refractor of 1824 (see Chapter 11). Other large instruments followed, like the two 15 inch (38 cm) telescopes built for Pulkowa in Russia and Harvard in the USA during the early 1840s by Fraunhofer's successor, Georg Merz (1793–1867), and his partner Franz Joseph Mahler (1795–1845). These telescopes were intended primarily for planetary and double-star measurements, and had little to do with the ongoing problem of the nature of nebulae. That obsession was unquestionably the province of the large-aperture reflector, exemplified by Lord Rosse's Leviathan and the Great Melbourne Telescope.

Or was it? As it turned out, it was a modest refracting telescope in the hands of a modest amateur astronomer that proved just how wide of the mark this idea was. Moreover, the dramatic demonstration took place even before the order for the Great Melbourne Telescope had been placed, highlighting once again the impotence of its design.

The refracting telescope in question was an 8 inch (20 cm) equatorial built by Thomas Cooke (1807–1868) of York, perhaps England's foremost instrument-maker of the time. It had been ordered in 1858 by one William Huggins—the amateur in question—to accommodate an objective lens he had just bought second-hand for £200. Though that was a king's ransom in those days, it probably seemed like a bargain to Huggins. The objective was of exquisite quality. It had been made by the American optician Alvan Clark (1804–1887), one of that country's first successful telescope lens-makers.

Huggins himself was a man of independent means. Born in 1824, he had sold the family business in the mid-1850s and devoted himself to astronomy at his new home in the (then) dark skies of Tulse Hill, South London. Having taken delivery of the new telescope and explored its capabilities with the usual celestial sightseeing, he started to wonder what serious work he could do with it. The answer—which was soon forthcoming—was to pioneer the technique of stellar spectroscopy. It was William Huggins who perfected that marvellous celestial conjuring trick of splitting starlight into its component rainbow colours to reveal intimate stellar secrets across the abyss of space. In Huggins' time, this was cutting-edge astronomy. Remarkably, it still is today.

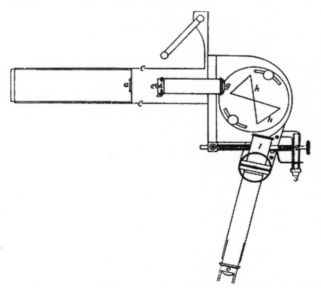

William Huggins' first spectroscope for investigating the composition of the stars, 1862.

Astronomers had been looking at the spectrum of one particular star since the seventeenth century. Isaac Newton was only 29 when he wrote:

> I procured me a Triangular glass-Prisme, to try therewith the celebrated *Phaenomena* of *Colours*. And in order thereto having darkened my chamber, and made a small hole in my window-shuts, to let in a convenient quantity of the Suns light, I placed my Prisme at its entrance, that it might be thereby refracted to the opposite wall. It was at first a very pleasing divertisement, to view the vivid and intense colours produced thereby . . .

Newton's discovery that white light is composed of individual spectrum colours was only properly understood when, at the turn of the nineteenth century, Thomas Young (1773–1829) recognised that the colours correspond to light of differing wavelengths.

And then, in 1802, William Wollaston (1766–1828) noticed mysterious dark lines crossing the solar spectrum—although he rather missed the point when he took them simply to be the boundaries between one colour and the next. During the ensuing sixty years, with the work of Fraunhofer, Gustav Kirchhoff (1824–1887), Robert Bunsen (1811–1899) and others, the true significance of the dark lines as chemical messengers from the Sun's atmosphere was recognised. They were 'absorption lines', dark lines whose position in the solar spectrum coincided exactly with the positions of bright 'emission lines' (light of a single wavelength) in the spectra of various terrestrial elements such as calcium or iron when they were excited by a spark or flame in the laboratory (see 'Light with a bar code', opposite). It was a gigantic step forward, and it paved the way for the birth of astrophysics—the physics of the stars.

Light with a Bar Code

It is an instrument known as the spectroscope that allows us to see light spread out into rainbow colours, from deep violet to deep red. This range of colours corresponds to the spread of wavelengths that makes up visible light, from the shortest to the longest. Mixed together, they form plain white light. The science of spectroscopy therefore allows us to analyse or dissect the component colours of light.

When the spectroscope is pointed at an ordinary lamp-bulb filament (which glows simply because it is being heated), the spectrum has the appearance of a continuous band, known, not surprisingly as a *continuous spectrum*. But gases made to glow by passing an electrical current through them, or by being burned in a flame, produce a very different spectrum—an *emission-line spectrum*—which contains only certain colours (corresponding to specific wavelengths). These appear as bright lines across the dark background where the rest of the continuous spectrum would be if a lamp-bulb were being observed. Crucially, the set of specific wavelengths emitted—and therefore the pattern of lines observed—is quite different for different gases. The light emitted by glowing gases is therefore imprinted with a kind of bar code that tells the spectroscopist exactly what material is being observed.

Another vital aspect of spectrum analysis is that if a hot object is viewed through a gas at a lower temperature, then the continuous spectrum is crossed by dark lines that are exactly where the emission lines of the gas would appear if it were glowing. Such an *absorption-line spectrum* is exactly what is observed in stars like the Sun, where the hot surface is observed through the cooler gas of the star's atmosphere.

Fraunhofer's name is immortalised in the Fraunhofer lines—the name given to the solar absorption lines today— and he also seems to have been the first person to observe the spectrum of a star other than the Sun (Sirius, in 1814). But it is Huggins' name that is always associated with the transformation of this colourful novelty into a vital diagnostic tool.

In 1862, just as Huggins was wondering what to do with his splendid new telescope, he heard of Kirchhoff's detailed analysis of the solar spectrum.

> This news was to me like the coming upon a spring of water in a dry and thirsty land [he wrote in 1897]. Here at last presented itself the very order of work for which in an indefinite way I was looking— namely, to extend [Kirchhoff's] novel methods of research upon the sun to other heavenly bodies.

Huggins promptly fitted his telescope with a two-prism spectroscope built with the help of his friend William Miller (1817–1870), a professor of chemistry at King's College. Together, the two men embarked on an out-and-out romp through the heavens, observing the spectra of the Sun, Moon, planets—and, most significantly, the stars.

Of course, it was only the brighter stars that yielded their rainbow secrets to the crude equipment of the two pioneers. But such was their perseverance and enthusiasm that in 1864, Huggins and Miller were able to present ground-breaking results on the spectra of some 50 stars, unambiguously identifying the absorption lines they had found with emission lines in the flame spectra of terrestrial elements. As Huggins later wrote:

> One important object of this original spectroscopic investigation of the light of the stars and other celestial bodies, namely to discover whether the same chemical elements as those of our earth are

present throughout the universe, was most satisfactorily settled in
the affirmative; a common chemistry, it was shown, exists throughout the universe.

The new science of astrophysics was on its way.

It was late in the summer of 1864 that Huggins made his
most spectacular discovery. And, yes, it concerned the nature
of nebulae. Following Lord Rosse's observations in the 1840s
(see Chapter 12), many astronomers considered *all* these fuzzy
objects to be aggregations of stars too distant to be resolved
into individual points of light. Indeed, that is what some of
them eventually turned out to be, and today we call the biggest
of those particular specimens galaxies.

But what were the symmetrical objects that, because of
their superficial resemblance to planets, William Herschel had
christened 'planetary nebulae' in 1785? Herschel himself,
having discovered a bright star at the exact centre of one such
nebula in 1790, was convinced they were not made of stars—
but could offer no plausible alternative. Then, in 1864, along
came Huggins with his spectroscope.

On the evening of August 29th [he wrote in 1897] I directed the
telescope for the first time to a planetary nebula . . . I looked into
the spectroscope. No spectrum such as I expected! A single bright
line only! At first I suspected some displacement of the prism, and
that I was looking at a reflection of the illuminated slit from one of
its faces. This thought was scarcely more than momentary; then the
true interpretation flashed upon me. The light of the nebula was
monochromatic [i.e. an emission line] . . . The riddle of the nebulae
was solved. The answer, which had come to us in the light itself,
read: Not an aggregation of stars, but a luminous gas.

Huggins quickly followed up this remarkable observation
with similar studies of the Great Nebula in Orion and a

handful of other diffuse nebulae. In each case he found the same thing—bright emission lines from a luminous gas. While it took another sixty years or so to sort out the finer details of the composition of the gas (mostly hydrogen and oxygen), the riddle was, indeed, solved. Nebulae came in two sorts: those made of stars and those made of gas.

Thus was the enormous power of Huggins' technique demonstrated. With such a major discovery under his belt, he went from strength to strength, observing everything from comets to novae, and, in the process, pioneering the use of photography as a tool in astronomical spectroscopy. He was supported throughout by his wife, Margaret (herself an accomplished amateur astronomer), and no doubt was as delighted as she was when she found herself transformed from plain Mrs 'Uggins to Lady Huggins on his knighthood in 1897. William became a major figure in the Royal Society, serving as its president from 1900 to 1905. By the time of his death in 1910, the whole world of astronomy had accepted the vital role of spectroscopy in opening up the physics of the stars. As he himself had written as early as 1866:

> So unexpected and important are the results of the application of spectrum analysis to the objects in the heavens, that this method of observation may be said to have created a new and distinct branch of astronomical science.

He was quite right.

RECORD BREAKERS

It was the combination of an inspired new technique and the high light-efficiency of a refracting telescope that had allowed

William Huggins to make his great breakthrough in the study of nebulae. And, while the discovery might not have influenced the fortunes of the large refractor directly, it certainly did them no harm.

The progression from one large refracting telescope to another even larger one continued steadily as the century progressed. In 1870, Thomas Cooke posthumously took the record for the world's largest refractor when the 25 inch (63.5 cm) telescope he had been building since 1862 for a Gateshead amateur astronomer, Robert Stirling Newall (1812–1889), was delivered. Its career in the north-east of England was unspectacular, however, and in 1890 it was transferred to Cambridge University. Eventually, it was moved again, this time to Greece, where today it performs useful service at the National Observatory of Athens.

The Newall Telescope's moment of glory as a record-breaker was short, for in 1872, Alvan Clark completed a 26 inch (66 cm) refractor for the US Naval Observatory in Washington, DC. Clark's rise to telescope-building stardom had been interrupted by the American Civil War of 1861–65, but now he was able to deliver objective lenses of the same exquisite quality as Huggins' 8 inch lens—but in world-beating diameters. It was with this instrument that the American astronomer Asaph Hall (1829–1907) discovered the two tiny moons of Mars, Phobos and Deimos, in August 1877.

Another six years were to pass before the record fell again. This time, the winning manufacturer was none other than Howard Grubb, the Dublin instrument-maker who had cut his teeth on the Great Melbourne Telescope. The new telescope was a 27 inch (68.5 cm) refractor for the Vienna Observatory, ordered in June 1875. Its lenses were to be made from glass supplied by Charles Feil in Paris, a grandson of Pierre

Louis Guinand (see Chapter 11). Unfortunately, the crown glass disc took several attempts to manufacture, and the completed telescope was not delivered until mid-1883. It was, however, a triumph, and Grubb's reputation soared as a result. He was duly knighted in August 1887. Sadly, old Thomas Grubb was not around to see his son's great achievement, for he had died in 1878.

During the second half of the 1870s, Howard Grubb engaged in a lengthy correspondence with one Richard S. Floyd (1844–1891), head of the Board of Trust of what was eventually to become the Lick Observatory. James Lick was a highly eccentric Californian millionaire who wanted to ensure that his name would live on in perpetuity. He had been persuaded that the best way to do this would be to endow a great telescope, and the wheels to that end were duly set in motion.

James Lick died on 1 October 1876 at the age of 80, and that melancholy event prompted an immediate crisis in the telescope project. His son, John, born out of wedlock to Lick's sweetheart Barbara Snavely in 1817, contested the will. Since John had been cut out for the sole reason that he had neglected a pet parrot, he probably felt entirely justified in claiming a large share of the estate on the grounds of his late father's insanity. His suit was unsuccessful.

Despite the litigation the Board of Trust pressed on and eventually, in December 1880, a contract for a 36 inch (91 cm) objective lens was let—not to Grubb, but to Alvan Clark. Progress then ground to a halt again while the Feil company struggled to make the gigantic glass discs needed for such a lens. In the event, it took no less than *twenty* attempts to cast the crown glass component.

Meanwhile, Howard Grubb continued to plan the telescope and dome—which would incorporate his novel invention of

Grubb's design for the Lick telescope.

a rising floor so that the eyepiece would always be at a convenient height for the observer, no matter where the telescope was
pointing. When, in 1886, the contract was given instead to the
relatively new Cleveland, Ohio, firm of Warner and Swasey
(founded in 1880), Grubb was understandably disappointed.
The financial reimbursement he got for his lifting-floor idea
was small compensation for the effort he had put into the
project.

The Lick Telescope finally came into use in 1888 at what
was the first mountaintop observing site in the United States.
Its home on Mount Hamilton is some 32 kilometres east of
San Jose, California, at an altitude of 1280 metres (4200 ft).
It remains the world's second-largest refractor. From the

beginning, the telescope was operated by the University of California. And since its beginning, no observer has ever used the instrument alone. There is always silent company in the form of the observatory's eccentric benefactor, whose remains lie in the final resting-place he chose before his death—underneath the telescope's support pier.

Only two refractors have surpassed the Lick 36 inch in size. The first was the Yerkes 40 inch (102 cm) telescope, which was completed in 1897 and is today the world's largest. It owes its origin to an entrepreneurial solar astronomer named George Ellery Hale (1868–1938), who cajoled the streetcar magnate Charles Tyson Yerkes (1837–1905) into financing an observatory for the University of Chicago. Like the 36 inch, the telescope was built by the combination of Alvan Clark and Warner and Swasey, but this time with glass lens blanks from Mantois of Paris (another of Guinand's successors). Mounted in their supporting cell, the finished lenses weigh half a tonne, demanding the utmost rigidity from the 62 ft (18.9 m) long tube.

The Yerkes refractor has a rising floor based on Grubb's design for Lick. It weighs almost 37 tonnes and, a few days before the telescope's opening ceremony, it made a spectacular bid for freedom by slipping from one of its supporting cables and crashing 45 ft (13.7 m) to the ground. Fortunately, no one was injured in the accident, although the damage took some time to repair. This was, however, the only black mark in a highly productive career that contributed spectroscopic, photographic and, eventually, electronic observations to almost every branch of astronomy.

In stark contrast, the largest refractor ever made did not produce a single discovery. This ill-fated instrument, known

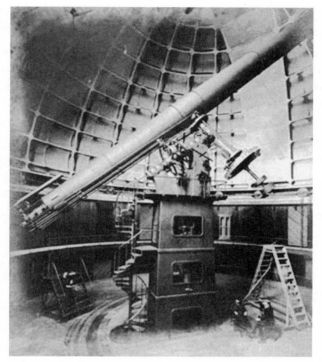

The Lick 36 inch refractor. Grubb's tender was unsuccessful, but features like his rising floor were incorporated into the final 1886 design.

as *La Grande Lunette de 1900*, or the Great Paris Exhibition Telescope, was a superb piece of optical engineering that unfortunately found itself in the wrong place at the wrong time. It was funded by the float of a public company, *L'Optique, Société Anonyme des grands télescopes*, which existed purely to build the instrument and display it in working order to the public at the Paris *Exposition Universelle*, beginning with the exhibition's opening on 14 April 1900.

The telescope used lenses 1.2 m (4 ft) in diameter with a focal length of 57 m (187 ft)—three times the length of the Yerkes refractor. To allow public access to such a colossal instrument, it was mounted with the tube permanently horizontal. Starlight was fed to it by means of a 'siderostat', which used a 2 m (6ft 6in) diameter flat mirror on a complex equatorial mounting to reflect any selected area of sky into the tube. The 2.2 tonne glass mirror was itself a major triumph, and is preserved today at the Paris Observatory.

Even more interesting were the objective lenses, polished from discs cast, like those for Yerkes, by Mantois. The plan called for two, each an achromatic doublet, with one corrected for visual use and the other for photography. (They were different because of the sensitivity of turn-of-the-century photographic plates to blue light, compared with the eye's peak sensitivity to yellow-green.) The two objectives were to be mounted side by side—rather like a gigantic pair of spectacles—and interchanged by means of a sliding mechanism. In the event, only the photographic one was completed and used. For many years after the exhibition, its component lenses were thought to have been lost, but French astronomers were recently delighted to learn that they have been rediscovered, gathering dust in a cellar at the Paris Observatory.

Details of the life of this telescope are sketchy, but it seems that it was used successfully with the photographic objective during the exhibition—to the extent that the skies of central Paris would allow useful observation. But the operating company, *L'Optique*, folded at the end of the public display, and no source of funds could be found to re-erect the telescope on a dark-sky site for professional use. Consequently, it was scrapped, a sad end for an ambitious and potentially great instrument. The largest refractor in Europe today remains the 83 cm (33 inch) *Grande Lunette*, built a few years before the Paris telescope by the same manufacturer, Paul Gautier (1842–1909), and located at nearby Meudon.

As the twentieth century dawned and flourished into the Edwardian era, it soon became clear that the large refracting telescope was about to be overtaken by the silver-on-glass reflector. Perhaps that was a source of frustration for the burgeoning German optical industry, with its newfound expertise in the production of specialised lenses. But, in the event, there was no shortage of customers for its products.

As diplomatic tensions grew throughout Europe, novel optical instruments began to be manufactured in great numbers: rangefinders, stereo-telescopes, field periscopes, trench binoculars, gunsights, panoramic telescopes and ordinary service binoculars. In the design and production of such equipment, Germany reigned supreme.

When political brinkmanship cascaded into the slaughter of the Great War, these optical munitions quickly took on strategic importance. It is a measure of just how unprepared the rest of the world was when, in 1915, the British Government quietly attempted to barter with the enemy to obtain binoculars and telescopic gunsights. But such opportunism was not one-sided. As increasingly sophisticated optical instruments faced one another in their scores across the muddy wastes of no-man's-land, the dream optics of Abbe and Zeiss joined with the strains of Brahms' sublime *Requiem* to become unintended icons of German nationalism.

And the science of the stars became but a memory.

15

SILVER AND GLASS

THE TWENTIETH-CENTURY TELESCOPE

Monsieur Foucault was not amused. 'Le télescope de Lord Rosse est un blague,' he fumed. 'Pour les anglais le mien n'existe pas . . .'.

> Lord Rosse's telescope is a joke. For the English, mine does not exist: it has been, it is, it will be for some time as if it had never happened; in spite of it they made me an honorary doctor.

Written after a visit to Ireland in 1857, Foucault's words were understandably bitter—even if he had rather got hold of the wrong end of the stick regarding the exact nationality of his hosts at Dublin University.

Léon Foucault (1819–1868) was a physicist who, in that same year of 1857, had completed a refined 33 cm (13 inch) aperture reflecting telescope on a sturdy fork-type equatorial mounting. Refined it might have been, but surely, if it were compared with Rosse's 6 ft Leviathan, wouldn't the joke be on Foucault? But Foucault's little telescope boasted something new. At its heart was a mirror quite unlike the heavy lumps of speculum metal that had been used since Newton's time. And in its bright reflections lay the future of the telescope.

Foucault was not the first optician to use a silvered-glass mirror. A year earlier, the Munich instrument-maker Carl August von Steinheil (1801–1870) had produced a 10 cm (4 inch) telescope that had clearly demonstrated the effectiveness of a fine layer of silver deposited chemically on to a glass surface. But it was Foucault who was the visionary and, by 1862, he had completed an 80 cm (31 inch) telescope that finally established this new technology as a viable alternative to speculum metal.

In fact, the new technology won hands down. A glass mirror was less than one-third the weight of its metal equivalent. Its silvered surface was so much more reflective that a Newtonian or Cassegrain telescope would gather *twice* as much light as a metal-mirrored instrument of the same size. And, even if the silver did tarnish over time, it was a straightforward matter to recoat it chemically—no need for risky optical polishing.

In spite of all that, the old guard resisted its introduction— particularly in the Great Melbourne Telescope, as we saw in Chapter 13. It was not until 1872 that the Melbourne's maker, Howard Grubb, completed his first silver-on-glass mirror—a 24 inch (61 cm) for Scotland's eccentric, pyramid-obsessed Astronomer Royal, Charles Piazzi Smyth (1819–1900).

Foucault made another crucial breakthrough in telescope-making. In 1859, he devised a simple test that would allow an optical worker to map out the errors in a mirror's surface in great detail. This 'knife-edge' test (so called because it literally uses a knife-edge to cut the rays of light) paved the way for mirrors of extremely high definition to be produced. Once the errors had been accurately charted, they could be corrected by judicial polishing. It opened up the possibility of real competition with the best objective lenses.

There was yet another way in which mirrors could now compete with the lenses of refracting telescopes. Glass, being an elastic material, has a tendency to sag under its own weight. In ordinary windows the effect is imperceptible, but in a lens whose surface must be accurate to a tiny fraction of a millimetre it starts to become a problem when the diameter of the unsupported glass approaches a metre or so. A mirror, on the other hand, can be supported from the back as well as the edge, using a flotation system like the ones developed by Thomas Grubb and William Lassell (see Chapter 12). There is consequently no real limit to the size of a mirror, whereas the maximum diameter for edge-supported lenses is about the size of the Yerkes and Paris objectives. It was this simple fact more than anything else that set the stage for the reflecting telescope's spectacular rise to supremacy during the twentieth century.

Other issues were also starting to become important. Following pioneering studies by Warren De la Rue (1815–1889) and Andrew Common (1841–1903) in Britain, and the Americans Henry Draper (1837–1882) and Lewis Rutherfurd (1816–1892), photography had become a major observing tool by the mid-1880s. Like the as-yet-undreamed-of electronic detectors that would supplant them a century later, photographic plates had two major advantages over visual observation.

The first was simply that they provided an accurate and permanent record of what was in the sky. This aspect made such a deep impression on David Gill (1843–1914) at the Cape, and the Henry brothers (Paul, 1848–1905, and Prosper,

1849–1903) in France, that in 1887, these gentlemen became the driving force behind a massive international collaboration to map the entire sky photographically. This ambitious project—the *Carte du Ciel*—involved no fewer than eighteen observatories. It was never properly completed, although its associated star catalogue was published in 1962.

The other advantage of photographic observation over visual work was even more significant. For all its remarkable sensitivity, a dark-adapted human eye can only perceive what is apparent in a single glance. But a photographic plate or film can gradually build up an image over a period of time, recording faint details that would never be visible to the eye, no matter how long its owner stared through the eyepiece. As long as light continues to fall in exactly the same spot on a photographic plate, information will continue to be added to the image.

This immediately placed new demands on the mechanisms used to keep equatorial telescopes pointing at the same area of sky as the Earth turned beneath them. Photographs requiring many hours of exposure called for the utmost accuracy in tracking. But photography's huge potential for revealing faint celestial objects also had a profound effect on the design of telescopes themselves. As the nineteenth century gave way to the turbulent twentieth, the reflecting telescope began its gradual metamorphosis from a slender tube into the squat giants that are today's icons of optical astronomy.

ABSOLUTELY NEBULOUS

The architect of that process was a gifted American telescope-builder and astronomer called George Willis Ritchey

(1864–1945). In 1901, Ritchey was head of instrument construction at the Yerkes Observatory of the University of Chicago, and he had just completed a 23.5 inch (60 cm) reflecting telescope. Remarkably, this modest instrument was soon providing real competition for the observatory's giant 40 inch (102 cm) refractor, the largest lens telescope in the world.

Ritchey was interested in astronomical photography and, in particular, in recording images of nebulae. No surprises there—the mystery of the true nature of spiral nebulae was still the hot topic of the day. But Ritchey was aware of a curious and recently discovered (1882) attribute of photographic image detection—namely, that the speed with which an extended object like a nebula or a comet can be recorded depends not on the aperture of the telescope but only on its focal ratio, or f/number. This quantity, well known to photographers, is just the focal length of the lens or mirror divided by its diameter.

It seems almost impossible to believe that sensitivity could be independent of aperture, but it is true—the lower the focal ratio (and therefore the stubbier the telescope), the greater the speed with which the image of a nebula will be detected. That is why low focal-ratio lenses or mirrors are often called 'fast'. There are still very good reasons for making them as large as possible, however, for their diameter determines both the amount of detail that is recorded and their sensitivity to point-like sources such as stars.

Ritchey's new 23.5 inch telescope had a mirror with a focal ratio of f/3.9, astonishingly fast by the standards of the day. By contrast, the 40 inch worked at f/18.6, which was slow even compared with other late nineteenth-century refractors. Little wonder the new reflector could beat it hands-down for nebular photography.

The early decades of the twentieth century saw Ritchey's work on large reflecting telescopes flourish, mainly because of his collaboration with the visionary astronomer George Ellery Hale. When Hale procured a 60 inch (1.52 m) glass mirror blank from the St Gobain glassworks in Paris with funds from his father, Ritchey set about designing the instrument that would accommodate it. By 1908, it had emerged as the most sophisticated telescope ever built, with a choice of no less than four different secondary mirror arrangements for various purposes, and a fork mounting whose polar axis floated on mercury to minimise friction. It spelled 'the definitive death-knell for the large refractor', as one modern commentator has put it.

But the 60 inch was not built at Yerkes. Lack of funding and a subsequent rescue bid by the Carnegie Institution of Washington meant that it ended up atop a 1740 metre (5700 ft) mountain near Pasadena in southern California. And as a result, the Mount Wilson Observatory—founded in 1904 as a solar observatory by Hale himself, and still operated today by Carnegie—soon drew the attention of the world's astronomers. By 1917, their attentiveness had been transformed into gasps of admiration as the 60 inch was followed by Ritchey's masterpiece—the 100 inch (2.54 m) Hooker Telescope. This great reflector, for thirty years the largest telescope in the world, holds a special place in the affections of astronomers. For it was with this instrument that the riddle of the spiral nebulae was, at long last, unequivocally solved, and our modern picture of a vast Universe populated by gigantic star systems began to emerge.

John D. Hooker (1837–1910) was the wealthy Los Angeles businessman who was persuaded by Hale to finance the telescope. His first instalment procured the 4.5 tonne mirror blank, again from St Gobain, which was delivered to the observatory's sea-level base at Pasadena in 1908. However, it was not until 1910 that Ritchey began the process of grinding and polishing the giant glass disc, a task that took no less than five years. The finished telescope was mounted on a modification of the English equatorial arrangement (Chapter 11), with the polar axis now taking the form of a rectangular steel cradle. In this so-called yoke mounting, the telescope swings between the longer sides of the cradle to provide the declination movement. It has the disadvantage that the celestial pole and its immediate surroundings are inaccessible—an indication of the engineering pragmatism necessary in this first-ever 100 tonne telescope.

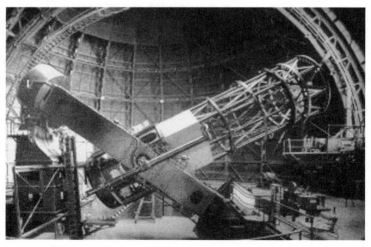

100 inch mirror, 100 tonne telescope. The Hooker Telescope at Mount Wilson has been a favourite with astronomers since its completion in 1917.

Like the 60 inch, the Hooker Telescope had a primary mirror focal-ratio of about f/5. This was somewhat slower than Ritchey's earlier 23.5 inch telescope—probably because he was aware of the difficulties in making fast mirror surfaces in larger sizes. Nevertheless, it was fast enough to reveal details in photographs of spiral nebulae, and it was with such images that a young Mount Wilson astronomer called Edwin Powell Hubble (1889–1953) made the Big Breakthrough in nebular astronomy.

Using a series of photographs taken with the Hooker Telescope between 1919 and 1924, Hubble found a certain type of periodic variable star (one whose brightness changes with time in a regular pattern) in two bright spiral nebulae in the northern constellations of Andromeda and Triangulum. These stars—known as Cepheid variables—have the property that their intrinsic brightness can be determined from the time-period of their brightness variation. Thus, they act as 'standard candles'—beacons of known luminosity.

The spiral nebulae, he discovered, were very remote. They were at distances measured in millions of light-years, placing them well beyond the bounds of our own Milky Way system of stars. They confirmed the idea that spiral nebulae were 'island universes'—immense star systems in their own right. Today we call them galaxies, and we also know that in the grand scheme of things, the Andromeda and Triangulum nebulae are *very* local objects.

But there was an immediate problem. Hubble's discovery flew in the face of results that had recently been obtained by another senior member of Mount Wilson's staff, a man by the name of Adriaan van Maanen (1884–1946). This Dutch-American astronomer had observations of spiral nebulae consisting of photographic plates taken by George Ritchey soon after the 60 inch telescope had been completed in 1908, together with a series of his own plates taken from 1915 onwards. When he compared the earlier plates with the later ones, van Maanen thought he could see a rotation of the nebulae—which could only be possible if they were small, nearby objects. Had they been large objects at great distances, their outer regions would have had to be flying around faster than the speed of light, which was already known to be unattainable.

Here, then, was an impasse—and an embarrassing one for Mount Wilson Observatory, anxious to present a united front to the scientific world. The person in whose lap the problem fell was the observatory's director, Walter Sydney Adams Jr (1876–1956). He exercised considerable behind-the-scenes skill in engineering the eventual outcome—two adjacent papers in Vol. 81 (1935) of the *Astrophysical Journal* presenting the views of each protagonist in a mutually acceptable way, and one that did not undermine the integrity of the observatory.

What has only recently come to light, in Mount Wilson archives made public in the early 1990s, is the struggle Adams had on his hands to achieve that outcome. It seems that once Hubble had discovered the true distances of spirals, he wanted to get to the bottom of the inconsistency with van Maanen's observations, so he and two other astronomers re-measured the 60 inch photographs. They could not repeat the results. It now appears likely that the earlier plates were actually test photos made when the adjustment of the telescope was incomplete, and the inferior quality of the images was what gave rise to the erroneous result. Van Maanen, however, was convinced that the rotation was real—and stuck to his opinion.

At this point, Hubble embarked on a campaign of vilification against van Maanen, writing a document for publication containing language that was 'intemperate in places', and showing a marked 'attitude of animosity'—as Adams put it in a confidential 1935 memorandum. Hubble was clearly angry at the baseless nature of van Maanen's result—and with van Maanen himself for sticking to it. He would not give an inch in conciliation and, as Adams said:

> The situation then became very difficult ... The attitude of van Maanen in the matter was much superior to that of Hubble: van Maanen, who fully believes in the existence of the motions indicated by his measures, went far in acknowledging the probable existence of systematic errors; while Hubble, who had much the better of the general weight of evidence, showed a distinctly ungenerous and almost vindictive spirit. This is not the first case in which Hubble has seriously injured himself in the opinion of scientific men by the intemperate and intolerant way in which he has expressed himself.

It is a tribute to Walter Adams' diplomacy that he eventually managed to get Hubble to withdraw his 'intemperate' document, and rewrite it as the brief note that appeared alongside van Maanen's in the *Astrophysical Journal*. Today, decades later, it is only Hubble's brilliance that we remember—not his vindictiveness. The fact that astronomy's most famous orbiting telescope is named after him simply underlines the fact that you don't have to be nice to become immortal.

WIDER PERSPECTIVE

For all that they spelled the end of the era of large refractors, Ritchey's new reflecting telescopes had an Achilles heel of their own—and it lay in the profile of their giant dished mirrors. As we saw in Chapters 7 and 8, a telescope mirror must have the subtle shape of a paraboloid to form a perfect point image when parallel light (for example, from a star) falls upon it. But that perfection is only achieved if the incoming light is exactly aligned with the axis of the mirror. Departure from this situation causes the images to be degraded by an optical aberration known as 'coma', which gives stars the appearance of little comets with their tails pointing away from the centre of the field of view—hence the name.

With a paraboloidal mirror, coma is noticeable only a fraction of a degree from the field centre. Even worse, it becomes more acute as the focal ratio gets faster. Thus, for all the potency of Ritchey's fast reflecting telescopes in nebular astronomy, they were almost useless for photography over very wide angles of view—which is exactly what was needed to record large, faint nebulae.

It was another eccentric, an ill-adjusted genius named Bernhard Voldemar Schmidt, who solved the problem of wide-angle photography with fast focal-ratio telescopes. This unusual man's life ranks among the most poignant episodes in the story of the telescope. He was born on the tiny Estonian island of Nargen in the Gulf of Finland on 30 March 1879, into a world that knew little but the day-to-day rituals of fishing, farming and the Lutheran Church. By the time of his death in Hamburg on 1 December 1935, he was celebrated as a master optician and brilliant innovator, but he had become deeply troubled by events in his adopted country. A pacifist, he foresaw the rise of Nazism leading inevitably to conflict and the military use of the invention he had cherished for its benefits to humanity. He was right: the Second World War saw both sides using Schmidt-type optics for strategic purposes.

What we know of Schmidt's early life on Nargen suggests

that he was an imaginative and adventurous youngster. On at least two occasions he came close to burning down the family home as a result of well-intentioned experiments that went wrong. More serious was the incident in which a metal tube packed with homemade gunpowder exploded in his hand, severing his right thumb and index finger. Horrific though the injury was, it was made much worse by the surgeon at the local Free Hospital, who could see no course other than to amputate Bernhard's whole forearm. Perhaps the resulting disability set him on course for the technical career he eventually pursued. Amazingly, it did not hinder his abilities as a practical optician, although he was always self-conscious about it. Only when the carnage of the First World War had made amputees a common sight did he feel somewhat less conspicuous.

Schmidt studied optics at the Chalmers Institute of Technology in Gothenburg, Sweden and then, at the turn of the century, moved to the small German town of Mittweida to continue his studies. He shunned the regimented formality of the academic environment, however, preferring to support himself by manufacturing small numbers of high-quality parabolic mirrors for both amateur and professional astronomers. In this, he became extraordinarily successful.

It was his skill as a practical optician that eventually brought Schmidt to the Hamburg Observatory in 1926, where he worked as a 'voluntary colleague'. This curious arrangement allowed the eccentric and solitary optician to come and go as he pleased, but still gave the astronomers at Hamburg access to his talents—if and when he felt like using them. Soon after his arrival, Schmidt met Walter Baade (1893–1960), one of the greatest astronomers of the mid-twentieth century. Baade encouraged Schmidt to explore the possibilities for designing a fast reflecting telescope with a truly wide field of view. When

Schmidt eventually succeeded in 1930, Baade was delighted, and it is due to him that the design received the recognition it deserved. Not only did he ensure that Schmidt's name was forever linked with the new instrument, but when he later moved to the USA, he championed the design among American astronomers. They, in turn, adopted it with enthusiasm.

What was this wonderful invention that so neatly solved the problem of recording faint nebulae over many square degrees of sky? In fact, it was simplicity itself, and the reasoning behind it showed Schmidt's crystal-clear understanding of the optical issues at stake. Noting that coma limits the usefulness of a paraboloidal mirror, he threw away the requirement for it to be paraboloidal and looked instead at spherical mirrors—ones whose cross-section is simply an arc of a circle. Such mirrors cannot normally be used in reflecting telescopes because of spherical aberration, that hoary old problem that prevents them forming sharp images of distant objects.

Schmidt realised that if the incoming beam were limited by a 'stop' (a hole, to you and me) placed at a spherical mirror's centre of curvature, there would be no preferred axis and therefore no coma. Only spherical aberration would remain—which, crucially, remains more or less the same at all image angles. Therefore, if you could correct the spherical aberration, you would have your perfect wide-field system.

And so, correct it he did. Schmidt's breakthrough was to replace the empty hole with a thin glass correcting plate having a shallow optical profile that would introduce into the incoming beam just enough spherical aberration to balance exactly that of the mirror. It was a brilliant solution, and it proved highly successful in practice.

Schmidt built a prototype telescope with an aperture (the

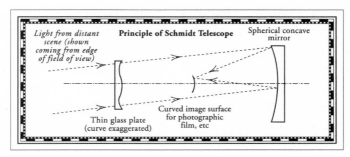

Light from distant scene (shown coming from edge of field of view)

Principle of Schmidt Telescope

Spherical concave mirror

Thin glass plate (curve exaggerated)

Curved image surface for photographic film, etc

The optical layout of a Schmidt telescope, incorporating a glass correcting plate and a spherically shaped mirror to gain a huge field of view.

diameter of the correcting plate) of 14 inches (36 cm). It had a focal ratio of f/1.7—extraordinarily fast—and its field of view was a staggering 15 degrees. And it worked superbly. At a stroke, Bernhard Schmidt had provided the world of astronomy with exactly the tool it needed to explore the sky at faint light-levels over wide areas. Sadly, he did not live to see the advances it eventually brought. Nor did his prototype survive for posterity. Like the peaceful aspirations of its gentle creator, it was destroyed during the Second World War.

Thanks to Baade's enthusiasm, Schmidt telescopes proliferated rapidly in the USA during the 1930s. But the design attained its most sublime form in the 48 inch (1.2 m) Palomar Schmidt (known today as the Oschin Schmidt), which was completed in 1948. It was built at Mount Palomar Observatory in California alongside Hale's masterpiece, the 200 inch (5.1 m) telescope that was for two and a half decades the world's largest. The symbiosis between these two instruments is legendary, the 'scouting' capabilities of the Schmidt allowing astronomers to forage for exciting new objects which could then be followed up in detail with the 200 inch reflector.

The Palomar Schmidt also engaged in photographic surveys of the whole northern sky, allowing astronomers free access to sky atlases of unprecedented detail and depth. In the wake of this success, other large instruments followed, and the period between 1948 and 1978 was truly the golden age of the photographic Schmidt telescope. No less than eight were built with corrector plates greater than 1 m (39 inches) in diameter. Because the plate is not a lens as such (for it is the mirror that does the focusing), flexure in the glass presented little difficulty compared with the lenses of refractors.

The largest southern hemisphere Schmidt is the 1.2 m United Kingdom Schmidt Telescope (UKST) in Australia, built in 1973 as a near-copy of the Palomar instrument. Both telescopes work at f/2.5, and were designed to take photographs of 6.6 degrees on a side on 14 inch (36 cm) square glass plates or films. That is big enough to allow the UKST to capture the entire Southern Cross in a single exposure, for example—a truly remarkable feat for a 1-metre class telescope. And each single photograph contains typically half a million faint stars and galaxies.

Like the Palomar (Oschin) Schmidt, the UKST was built close to a large conventional reflector, this time the 3.9 m (153 inch) Anglo-Australian Telescope (AAT) at Siding Spring Observatory in north-western New South Wales. Once again, this was no accident and the UKST also made complete photographic surveys of the southern sky. Today, however, neither of these Schmidts takes photographs. The Oschin Schmidt has been fitted with a gigantic battery of charge-coupled devices (CCDs)—electronic eyes far more sensitive than photographic plates that allow it to continue its role in deep, wide-angle imaging. But the UKST now works in a completely different way.

Back in 1982, two astronomers at the UKST proposed that optical fibres—flexible 'light-pipes' made of fine strands of glass—could be used to collect the light from many individual targets in the field of view and transmit them to a spectrograph for analysis. One of these scientists was John Dawe, then Astronomer-in-Charge at the UKST; the other was a character by the name of Fred Watson. The basic idea was not new: three years earlier, a group in the USA had demonstrated this multiple-object technique with the Steward Observatory's 2.3 m (90 inch) telescope at Kitt Peak National Observatory in Arizona. It had also been further developed at the Anglo-Australian Telescope. But Dawe and I were the first to point out the importance of field of view in using the technique—and its immense potential for Schmidt telescopes.

Twenty-odd years later, the UKST has become an instrument that spends all its time operating in this mode. What started as a crude experiment is now a sophisticated robotic process that allows the telescope to observe up to 150 celestial objects simultaneously. It is currently being used to make a three-dimensional map of 150 000 galaxies in the local Universe—a task that will be completed in 2005. Beyond that, it is hoped that a revolutionary new 2250-object system will allow the telescope to collect the rainbow spectra of some 30 million stars in our own Milky Way Galaxy. Such ambitious surveys bear eloquent testimony to the continuing effectiveness of Bernhard Schmidt's brilliant invention.

PALOMAR AND BEYOND

George Ellery Hale was a man gifted with extraordinary powers of persuasion. We catch a glimpse of them in an article he wrote in 1928 advocating the construction of a 200 inch (5.1 m) telescope:

> Starlight is falling on every square mile of the earth's surface, and the best we can do at present is to gather up and concentrate the rays that strike an area 100 inches in diameter.

And again, commenting on the enormous range in states of matter encountered by astrophysicists:

> A far-sighted industrial leader, whose success may depend in the long run on a complete knowledge of the nature of matter and its transformations, would hardly be willing to be limited by the feeble range of terrestrial furnaces. I can easily conceive of such a man adding a great telescope to the equipment of a laboratory for industrial research if the information he needed could not be obtained from existing observatories.

Far-fetched though that might sound to us in the economically rationalised twenty-first century, it was a ploy that worked. The result was a grant of $6 million from the Rockefeller Foundation to the California Institute of Technology to build and operate a 200 inch telescope.

As we have seen, that marvellous instrument eventually graced the summit plateau of Palomar Mountain in the San Jacinto range, 80 kilometres north of San Diego in southern California. Although its construction had started soon after the funding was granted—the mirror blank, for example, was cast in December 1934 after an unsuccessful earlier attempt—the project suffered inevitable delays during the Second World War. It was not until 3 June 1948 that the instrument was finally dedicated and named in honour of its great champion, George Ellery Hale, who sadly had not lived to see its completion. In its breathtakingly elegant dome which, at 42 m (137 ft) in diameter, remains one of the world's largest telescope enclosures, it quickly became an icon of post-war American science.

The Hale Telescope incorporated a number of important advances over its 100 inch predecessor. Its mirror, for example, was made not of plate glass like the Hooker's, but of Pyrex—a material developed by the Corning glassworks in the USA for ovenware. *Ovenware?* You might well ask what ovenware has to do with telescope mirrors. The answer is that both require a low rate of expansion with temperature, to minimise changes of shape. In ovenware, that means preventing the glass from splintering under heat, but in a telescope it means maintaining the perfection of the optical surface during ordinary day-to-day temperature changes.

The 100 inch Hooker Telescope had provided valuable lessons in this regard, and plate glass was now known to be quite unsatisfactory. A sudden change in temperature could render the instrument unusable for hours. Pyrex (actually a form of borosilicate glass) was far from perfect, but it was much better than plate. The mirror of the 200 inch was also lightened considerably by having a hexagonal pattern of ribs cast into its back; nonetheless, it still weighed some 15 tonnes.

Another improvement lay in the nature of the mirror's reflective coating. During the early 1930s, experiments had been carried out on the deposition of aluminium on to a mirror surface by evaporation. The process involved placing the mirror inside an airtight tank, pumping out all the air, then boiling a small quantity of aluminium in the vacuum. The end result was a thin, uniform coating of aluminium on everything inside the tank—including the mirror. Aluminium coatings are less prone to tarnishing than silver, and also reflect ultraviolet light more efficiently. After the Hooker Telescope's 100 inch mirror had been coated in this way in March 1935, vacuum deposition became the standard procedure for all astronomical telescopes—and remains so today.

With its f/3.3 mirror, open-tube structure and equatorial yoke mounting, the Hale Telescope represented the state of the art when full-time operations began in 1950. Its 500-tonne moving structure floated on thin films of oil under high pressure to minimise friction, for example. It had a huge open bearing shaped like a horseshoe at the northern end of its yoke to allow access to the celestial pole. It was fully equipped with spectrographs at its Cassegrain and other foci. And the coma of its parabolic mirror could be corrected with lenses near the mirror's focus, allowing half-degree diameter photographs to be taken. It is little wonder the telescope dominated optical astronomy for the next thirty-odd years, and still awes those scientists fortunate enough to win observing time on it.

By contrast, the instrument that eventually supplanted it as the world's largest turned out to be something of an under achiever. The 6.05 m (238 inch) Bolshoi Teleskop Azimutal'ny (BTA), or Large Altazimuth Telescope, was the former Soviet Union's Cold War response to the Hale. Located at the Special Astrophysical Observatory near Zelenchukskaya in the North Caucasus, it was disadvantaged primarily by poor imaging due to atmospheric turbulence. Although the telescope was not completed until 1976, its f/4 mirror was made from a material similar to Pyrex—which had by then long been superseded in the western world. The mirror's enormous mass of almost 43 tonnes compounded the problem of poor imaging, for it seldom achieved the even temperature needed to produce a perfect surface.

In one respect, however, the BTA was revolutionary. For the first time in a large, modern telescope, the time-honoured equatorial design involving a tilted polar axis was abandoned, and the instrument was steered around the sky with only the up-and-down and side-to-side motions of an altazimuth mounting. This greatly simplified the mechanical engineering at the expense of complexity in the control system, but since computers were now cheaper than huge horseshoe bearings, it was a very satisfactory trade-off. It set the pattern for almost all large telescopes completed since 1980, including today's generation of 8-metre to 10-metre class telescopes. The first of those giant instruments—the W.M. Keck Telescope, with its segmented 10 m aperture mirror—superseded the Russian facility as the world's largest in 1991. Located at the summit of Mauna Kea on the Big Island of Hawaii, the Keck was joined by its twin (Keck II) in 1996.

ON THE FACTORY FLOOR

Perhaps more important than the BTA in defining the archetypal telescope of the late twentieth century was a clutch of instruments in the so-called 4-metre class, built during the 1970s and early 1980s. No doubt their introduction also hastened the eradication of the old Imperial units from the vocabulary of telescope-building.

There were eight of them, with apertures ranging from 3.5 m to 4.2 m. They were sited in both hemispheres (five in the north and three in the south), and they imparted unprecedented momentum to optical astronomy on a truly global scale. Combined with a new generation of electronic detectors twenty times as sensitive as photographic plates, these instruments expanded humankind's horizons from millions of light-years to billions.

All but one of them (the 4.2 m William Herschel Telescope at La Palma in the Canary Islands) still used old-fashioned equatorial mountings. Some have criticised this as a legacy of conservatism left over from the Hale Telescope era. But without exception, the 4-metre class machines had mirrors made of advanced materials. Most used a cleverly contrived glass-ceramic mix that did not change its shape with temperature, making it perfect for telescope optics. This material was produced by Owens-Illinois in the USA under the tradename 'Cervit', but the manufacturing process eventually fell foul of the Illinois Pollution Control Board and had to be discontinued. Environmentally cleaner versions superseded it. The best known today is 'Zerodur', produced in Germany by the Schott company.

The 4-metre class telescopes were manufactured at various plants throughout the world, but one long-established firm played a major role in no less than three of them. A century earlier, Sir Howard Grubb's company in Dublin had been famous for the Great Melbourne Telescope and, as we saw in Chapter 14, went on to produce some of the nineteenth century's great refractors. By the early 1920s, however, the firm was in difficulties. It had been contracted to the British Government to make submarine periscopes during the First World War, and had been badly affected by the slump in the nation's optical industry that followed the end of hostilities. In January 1925, the company went into voluntary liquidation.

Rescue was close at hand, however, and it came from perhaps the most appropriate knight in shining armour imaginable. Sir Howard Grubb's father, Thomas, had been a close associate of Lord Rosse of Leviathan fame. Rosse's youngest son, Sir Charles Parsons (1854–1931), had in turn become famous as an engineer, having perfected the steam turbine in 1884. Since Charles was also interested in optics, what could be more fitting than these two sons joining together? Accordingly, in April 1925, the scientific journal *Nature* reported:

> A new company, trading as Sir Howard Grubb, Parsons & Co., has purchased from the liquidator the goodwill, drawings, and sundry plant and machinery of the firm, and workshops of up-to-date design are being erected at Heaton, Newcastle-on-Tyne, especially suitable for the building of large telescopes and observatory equipment. The advice and experience of Sir Howard Grubb will be at the disposal of the new company . . .

The new firm, which became universally known as Grubb Parsons, went on to produce many major telescopes, including no less than five 74 inch (1.9 m) reflectors between 1935 and 1956. Sir Howard himself passed away in 1931 at the age of 87.

In July 1950, the company was joined by a bright young Cambridge graduate by the name of David Scatcherd Brown, who demonstrated a gift for the design and testing of large telescope mirrors. It was not long before DSB (as he was known throughout the works) progressed to the job of Optical Manager, working in collaboration with a practical optician called David Sinden who managed the glass shops from 1962. These two gentlemen proved a formidable combination:

> There could not have been two more different types working together [wrote George Sisson, a former Managing Director of the firm], the one with a deep mathematical insight and ability to interpret obscure testing problems, the other with the instinctive feel for working glass, the hardness of the pitch, the construction of the polisher and methods of working.

Some five years later, this accomplished team was joined by a decidedly less accomplished young physicist fresh from the University of St Andrews, who worked with them for two years. That person was me—and DSB was my first boss.

With the 20/20 vision of hindsight, it is clear that I failed altogether to capitalise on this golden opportunity—despite David Brown's generous mentoring. But such paradoxes were commonplace at Grubb Parsons. For all that the company was deeply involved with high-tech products, it seemed to have its feet planted firmly in the nineteenth century. While David Brown was pioneering the use of computers to assist in the optical surfacing of large mirrors, for example, a man by the name of Big Jim MacKay spent his life tending cauldrons of boiling pitch for the polishers in a room that resembled an annexe of hell. Such incongruities were symptomatic of a firm trying desperately to drag itself into the twentieth century.

Eventually, Grubb Parsons failed to make the transition. In 1985, just 150 years after Thomas Grubb had completed his first large reflector, it closed its doors for the last time. But that was not before the firm had produced three of the best telescopes of the 4-metre era: the 3.9 m Anglo-Australian Telescope (1974), the 3.8 m UK Infrared Telescope (1978) and the 4.2 m William Herschel Telescope on the island of La Palma (1987). It also built the 1.2 m UK Schmidt Telescope. I have had the great privilege of observing with all these fine instruments, and now have the good fortune to be Astronomer-in-Charge of two of them.

By the time Grubb Parsons closed down, David Sinden had left to form his own highly successful optical company in Newcastle-upon-Tyne. But David Brown stayed until the end as Grubb Parsons' Technical Director, and then moved on to a research fellowship at the nearby University of Durham. It was a fitting appointment, given his enviable reputation in optics. But it was short-lived. In July 1987, this good-natured and unassuming man died at the age of only 59. Australian astronomer Ben Gascoigne paid tribute to his achievements:

> He was a past master in the art of figuring large [optical] elements for telescopes, with hardly an equal in his generation. He must certainly be ranked with Ritchey and other great opticians of the past, and is assured of a high place in the hierarchy of British instrumental astronomers.

The present generation of telescope-makers is much the poorer for the loss of David Brown's wisdom. No doubt he would also have had sage advice for the planners of the next generation of 30 metre to 100 metre optical telescopes. And the astonishing instruments that undoubtedly lie beyond them would have thrilled him to the core.

16

WALKING WITH GALAXIES

TOWARDS THE HALF-MILLENNIUM

During the second half of the twentieth century, humankind finally came face to face with the Universe at large. Hot on the heels of Edwin Hubble's discovery of the true nature of galaxies had come his recognition in 1929 that they are flying away from one another at breakneck speed— and the further away they are, the faster they are travelling. That, in turn, had led physicists to realise that we live in an expanding Universe, in which the very fabric of space is being stretched, carrying the galaxies along with it. It suggested the Universe had come into being in an explosive event that has been called the Big Bang ever since the great British physicist, Fred Hoyle (1915–2001), disparagingly christened it thus on a BBC radio programme in 1948. Hoyle himself was an ardent supporter of the opposing Steady State Theory, now regarded as untenable.

It is the science of cosmology that tells us about the origin and evolution of the Universe. The latest research indicates that the Big Bang occurred 13.7 billion years ago, with an uncertainty of only about 200 million years. Cosmology also tells us that there are perhaps 100 billion galaxies in the Universe, and that each contains roughly 100 billion stars. By a curious coincidence, 100 billion is also the approximate number of cells in a human brain.

As the twentieth century progressed, well-endowed minds grappled with new problems in cosmology. Mounting evidence suggested that there was more to the Universe than met the eye—or the telescope—and that some kind of hidden matter dominated its material content, rather than the visible stuff. Since the 1970s, for example, it has been known that galaxies rotate faster than they ought to if stars and gas are all that is holding them together. Without a halo of mysterious dark matter, they would simply fly apart.

During the 1990s, the situation became even more complicated. Observations of exploding stars at very great distances revealed that the expansion of the Universe, instead of slowing down over time because of the gradual braking effect of gravity, is actually speeding up. This led to the suggestion of a dark energy—a kind of springiness in space itself—whose nature was just as baffling as that of dark matter. Mystery upon mystery.

Such results as these came not just from optical telescopes like the Hale, the AAT and the Keck. For the other great advance that had propelled cosmology forward during the second half of the twentieth century was the opening of astronomers' eyes to all the wavebands of the electromagnetic spectrum. Gradually, from the end of the Second World War, the word 'telescope' came to mean something more than just a machine for gathering visible light. By the end of the century, it had been transformed beyond recognition.

It was William Herschel who had pioneered the idea of 'invisible' astronomy with his discovery of infrared radiation at the end of the eighteenth century. He noted that 'radiant heat [from the Sun] will at least partly, if not chiefly, consist, if I may be permitted the expression, of invisible light'.

When your name is William Herschel you are permitted any expression you like, but such favour was not granted to the next pioneer of invisible astronomy. His name was Karl Guthe Jansky (1905–1950), and he worked at the Bell Telephone Laboratories in Holmdel, New Jersey. When, in 1932, he discovered radio interference at a wavelength of 14.5 metres coming from the Milky Way, he was largely

Built on wheels taken from a Model T Ford, Jansky's 1932 antenna was the first radio telescope.

ignored by the scientific community. Jansky's antenna—a crude structure of wood and brass rotating on wheels purloined from a Model T Ford—was the first radio telescope.

It was followed in 1937 by the first steerable-dish type instrument, a 10 m (31 ft) diameter reflecting radio-telescope, built not by Jansky but by another enterprising American called Grote Reber (1911–2002). Reber, who spent his later years in Tasmania, was for almost a decade the world's sole radio astronomer, observing not just the Milky Way but a whole new breed of radio sources. It was only after the Second World War that others began to take up the new science. British Army radar, working at a wavelength of 4.2 metres, had inadvertently detected the Sun in the dark days of 1942, mistaking it at first for enemy jamming. This was followed up after the war, and anti-aircraft radar antennas were soon being beaten into ploughshares as rudimentary radio telescopes.

It was, in fact, the availability of huge quantities of war-surplus radio equipment that precipitated the rapid development of radio astronomy. At first, the purveyors of this new science were engineers, who were regarded with deep suspicion by their optical counterparts. In Australia, for example, the Director of the Commonwealth Observatory (today's Mount Stromlo Observatory) was asked in 1947 where he thought radio astronomy would be in ten years' time. 'Forgotten' was his acerbic reply.

Eventually, though, the two wavebands were seen as complementary, each providing uniquely different views of the same Universe. Radio observatories were established in Australia, the Netherlands and the United Kingdom, and such instruments as the 76 m (250 ft) dish at Jodrell Bank (completed in 1957) and the 64 m (210 ft) at Parkes (1961) quickly became icons of the new science. Like their optical counterparts, these great telescopes have had very long working lives.

Today's radio telescopes take the form of either the familiar single dish or an array of many independent telescopes that can be combined to synthesise a much larger one. The world's largest single dish is the 300 m (1000 ft) telescope at Arecibo in Puerto Rico (1963), whose fixed bowl is built into a natural depression in the ground. It probably represents an upper limit to the size of this kind of instrument. With arrays, however, the sky itself is the limit—literally, as some radio telescope arrays operate in space. Perhaps even more ambitious is the Square Kilometre Array, a ground-based instrument designed to collect faint radiation emitted in the so-called Dark Ages before the first galaxies were formed. Its one million square metre collecting area will synthesise a 1000 kilometre diameter radio telescope when it is built, perhaps in central Australia, in the second decade of the twenty-first century.

Unencumbered by problems of atmospheric turbulence, radio astronomers have used their arrays to pioneer novel methods of mapping fine detail in celestial objects, and a resolution of 0.001 arcseconds—a milli-arcsecond—is now commonplace. Ground-based optical astronomers can only look on with envy.

TELESCOPES IN SPACE

On 1 January 1956, Richard van der Riet Woolley (1906–1986) took office at the Royal Greenwich Observatory as the eleventh Astronomer Royal. The institution had itself only recently moved from light-polluted Greenwich to Herstmonceux Castle, in rural Sussex. Woolley epitomised the feet-on-the-ground optical astronomer, and he had made his views about the newfangled radio astronomy very clear in his previous position—as Director of the Commonwealth Observatory in Canberra.

On his arrival in Britain, he was asked by a journalist what he thought of the prospects for space research. His reply still echoed through the corridors of Herstmonceux Castle when I arrived there nearly two decades later—'Utter bilge!' Less than two years after this definitive proclamation from Britain's most senior astronomer, the world entered the space age with the launch of the unmanned Soviet satellite, Sputnik 1.

To be fair, Woolley's disparaging comment was probably a spur-of-the-moment quip rather than a considered judgement, for he more than most would have appreciated the benefits of a telescope in space. What he could not have known, however, was the extent to which radiation normally absorbed by the Earth's atmosphere would eventually shape our view of the Universe.

The first of the new wavebands to be explored were those at the extremities of visible light: ultraviolet and infrared. Astronomers had begun probing the sky at short infrared wavelengths (the 'nearer' end of the infrared spectrum) using ground-based instruments during the early 1960s, and such near infrared astronomy is today included in the capabilities of large optical telescopes. It was only with instruments carried above the Earth's atmosphere, however, that the full potential of infrared astronomy in penetrating cosmic dust clouds and revealing cool stars was realised.

A succession of space infrared telescopes and their counterparts in the ultraviolet waveband began to revolutionise our picture of the Universe during the 1970s and 1980s. Such acronyms as IRAS (Infrared Astronomy Satellite) and IUE (International Ultraviolet Explorer) became household names to astronomers during this period. At the same time, the very short wavelength radiations we know as X-rays and gamma rays were also being probed from space, revealing cosmic processes at the highest energies.

Sometimes, there were unexpected windfalls. In 1967, a series of US Air Force satellites designed to detect gamma-ray flashes from illegal nuclear weapons tests began to see frequent bursts of radiation from the sky. An analysis concluded in 1973 that they must be coming from deep space. More observations followed, but it was not until nearly a quarter of a century later that astronomers realised just how deep in space these mysterious gamma-ray bursts were occurring. In 1997, a succession of quick-response follow-up observations with optical telescopes revealed that they come from objects literally billions of light years away. Since some gamma-ray bursters outshine everything else in the Universe during their brief moments of glory, they must be releasing vast amounts of energy. They are now thought to originate in supermassive stars exploding at the ends of their short lives, a phenomenon that no longer occurs and is only visible to us because of the time it has taken for the radiation to cover vast tracts of space.

Today's space-borne telescopes are many and various. Perhaps the most prominent are NASA's four 'Great Observatories': the Hubble Space Telescope (launched 1990 for ultraviolet and visible-light observations), the Compton Gamma-Ray Observatory (launched 1991 and de-orbited 2000), the Chandra X-ray Observatory (launched 1999) and the Spitzer Space Telescope (launched 2003 for infrared observations). In addition there is a plethora of telescopes operated by various space agencies, and designed to do a wide variety of tasks—from monitoring the Sun in X-rays to mapping the remnants of the Big Bang itself. Astonishingly, the faint microwave echo of the birth of the Universe is still detectable as a 'cosmic wallpaper' beyond which we can never penetrate. And, as if these invisible radiations weren't enough, there is a whole range of ground- and space-based instrumentation designed to detect sub-atomic particles from space: cosmic rays, protons, neutrinos.

Truly, the modern astronomer is confronted with a bewildering whirl of data with which to decipher the Universe. It is now impossible for any one individual to keep abreast of it all. And so, in one of the first great collaborative enterprises of the third millennium, that maelstrom of information is being organised into something called the Global Virtual Observatory—a world-wide network of computers that will allow multi-wavelength data on any object in the sky to be drawn from the archives with a simple click of a mouse-key. It will be the enduring legacy of all the world's great telescopes, in space and on the ground.

It was in the midst of all the excitement of late twentieth-century astronomy that one more profound realisation dawned on the world of science. And this has turned into something extraordinarily useful to cosmologists. It is that the telescope, far from being a product of human ingenuity, is an invention of nature itself.

Most people know, of course, that eyes have much in common with telescopes—they are image-forming systems, and if their crystalline lenses had long enough focal lengths, they would give us a magnified view of the world. More useful to survival is the wide-angle view that normal eyes provide—and none more so than in the humble scallop, whose eyes bear an uncanny resemblance to miniature Schmidt telescopes. But nature also makes telescopes on a vastly greater scale.

During the 1960s, a new class of celestial object was discovered in a very successful collaboration between optical and radio astronomy. It was given the name quasar, an abbreviation of quasi-stellar radio source (something that looks like a star, but has strong radio emission). We now know that quasars are the wildly energetic cores of delinquent young galaxies, and are powered by supermassive black holes—exotic gravitational sink-holes in space from which not even light can escape. Quasars are extinct today and, like gamma-ray bursters, are seen only at huge distances corresponding to look-back times of billions of years.

In 1979, astronomers using a 2.1 m (84 inch) telescope at the Kitt Peak National Observatory in Arizona discovered a double quasar. It consisted of two objects with identical spectral signatures separated by some six arcseconds on the sky. How could such a juxtaposition of indistinguishable quasars occur by chance? It defied the laws of probability, and some suggested it might be a double image of the same object—a kind of mirage in space. Other examples followed. Then, in the mid-1980s, mysterious arcs of faint light were seen in the neighbourhood of giant clusters of galaxies. What were they? Broken-off bits of galaxy composed of billions of wandering stars? Or was something infinitely more subtle going on?

Astronomers very quickly realised that, yes, they were seeing the manifestations of an extraordinary prediction that had been made half a century earlier by Albert Einstein (1879–1955). The legendary physicist had suggested that if a massive object such as a galaxy—or a cluster of galaxies—lies exactly between ourselves and a much more distant object, then it will behave like a gigantic lens, focusing an image of the far-off object into a ring of light. It does this because gravity itself bends light-rays—a consequence of Einstein's General Theory of Relativity of 1915, which had been verified four years later during a total eclipse of the Sun. Such a phenomenon is known as a gravitational lens, and the image it forms is called an Einstein ring. If the alignment is not perfect, double images or incomplete arcs will be formed—exactly as had been seen by the astronomers. And if, as sometimes happens, the intervening galaxy or cluster of galaxies is too faint to be seen, these double images or arcs will have no obvious source of gravitational lensing.

Today, we know of many gravitational lenses producing a wide range of image types. Perfect Einstein rings are rare, but faint arcs of light are seen in profusion in deep images of distant galaxy clusters, and are of immense value to astronomers and cosmologists. They tell us first about the distribution of matter—both seen and unseen—in the intervening cluster of galaxies. That is because the way in which bright and dark matter concentrates in space affects the image of the distant object in its shape and position. Gravitational lensing is therefore a tool for exploring the properties of dark matter.

But perhaps more important is the effect of a gravitational lens in amplifying light from the distant object. Galaxies and quasars so remote from us that they would normally be invisible can be made detectable by the lensing effect of much nearer galaxies. It allows us to measure the rainbow spectra of these incredibly distant objects and probe the conditions that prevailed when the light left them, back in the childhood days of the Universe. The gravitational lens is behaving exactly like a telescope—albeit a crude one, because the images it produces are far from perfect. But in amplifying the light, it is performing the function of a telescope lens millions—sometimes hundreds of millions—of light-years in diameter.

In building instruments to explore the Universe, the telescope-makers and astronomers of history have merely been imitating what nature has done with the largest structures at its disposal. Though most of the fine people we have met in this book never knew it, they were truly walking with galaxies.

EPILOGUE

21.09.2108

Looking back now on the eve of the telescope's first half-millennium, it is easy to see where the breakthrough came. The discovery of the Earth-crossing asteroid 2041 FU in March 2041, and the realisation over the following months that there was a 99.9 per cent probability of this 1-kilometre object impacting Earth in April 2060, led to the most intensive period of space engineering the world had ever seen. By comparison, the Chinese lunar expeditions of 2009–21 and the establishment of the Mandela International Mars Base at Isidis Planitia in the mid-2030s had paled into insignificance.

The successful deflection of 2041 FU into a safe orbit with ion-beam propulsion units marked humankind's coming of age as a spacefaring species. It also highlighted the crucial role of the telescope in revealing the presence of this sinister, coal-black object. For it has to be said that during the first half of the twenty-first century, the fortunes of the astronomer's telescope had been decidedly mixed.

It was not so much the technology that had been the problem, for that had generally worked well. The James Webb Space Telescope, for example, launched in the early 2010s, had provided magnificent insight into the near infrared Universe with 20 milli-arcsecond resolution. And a suite of upward-pointing optical telescopes using liquid mirrors formed by rotating baths of mercury had proved outstandingly successful. The end product of Canadian research begun in the 1990s, the largest of these instruments became the first operational 30-metre class telescope when it was commissioned in 2015. It preceded the 100-metre OWL, with its multi-segment steerable mirror, by some four years.

Despite growing awareness throughout the world of the ills of light pollution—together with increasingly tight legislation attempting to control it—the viability of ground-based optical observatories with ever-larger telescopes became steadily more difficult to maintain. OWL, in particular, suffered from high sky-background levels. That, along with the continuing unreliability of its multi-conjugate adaptive optics system, led to the extraordinary decision by the European Southern Observatory to dismantle it and disband the organisation in 2033.

Likewise, the radio sky became so polluted with telecommunications traffic that the billion-dollar Square Kilometre Array, located in central Australia since 2014, had to be transported by the Chinese to their radio-quiet base on the far side of the Moon in 2029. Almost eighty years later, this venerable instrument is still providing useful data for a few highly targeted projects.

Pollution, however, was only part of the story of astronomy's decline during the early twenty-first century. More significantly, the science seemed to have become a victim of its own success. Cosmology, for so long the mainstay of observational astronomy, had become such a precise science by the 2020s that the remaining problems were merely a case of dotting the i's and crossing the t's. The big questions of dark matter and dark energy had been solved by 2012 in a famous coup that had upstaged all the major telescope projects. The 3.9 m Anglo-Australian Telescope, by now nearly forty years old but recently equipped with a new multi-object spectrograph whimsically named TIPKISS (The Instrument Previously Kept In Strictest Secrecy), had looked far enough into the Universe to provide definitive clues about the Big Unknowns. In doing that, it was following a well-established scientific tradition of cheap, fast-track experiments creaming off the important results before the expensively engineered versions could come onstream.

With cosmology largely a solved problem, the attention of astronomers turned to the other big question about the Universe—are we alone? Here, the issues were very different. Rather than looking back to a time when galaxies began evolving from the primordial gas clouds left over from the Big Bang, astronomers needed to look in much greater detail at the nearby stars of our own Milky Way Galaxy. Distances measured in billions of light-years suddenly dissolved into a stone's-throw of a few dozen light-years.

The new science of astrobiology—the study of life beyond the Earth—had begun to flourish in the 1990s. It had brought together a broad range of disciplines: astronomy, geophysics, chemistry, biology and others. Of course, the question of extra-terrestrial life in our own Solar System had been answered during the first couple of decades of the twenty-first century by unmanned space-probes going both to Mars and to some of the more promising satellites of the outer planets. Living bacteria found on Mars, and mucus-like organisms in the ice-covered ocean of Jupiter's moon Europa, proved that, yes, there was life elsewhere in the Solar System—but that humankind still represented its most highly developed form.

What was of much greater interest—and a much bigger challenge—was the quest for life on the planets of other stars. It had been known since the end of the twentieth century that so-called extra-solar planets were relatively commonplace. At least 10 per cent of stars within a few light-years of the Sun seemed to have Jupiter-sized objects around them, mostly in peculiar orbits that rendered them quite unlike the Solar System. Was there any room for Earth-like planets in these systems?

That was a question that had proved tantalisingly difficult to answer. The trick of using a large telescope to find a wobble in the parent star's motion, betraying the presence of a planet, only worked for Jupiter-sized objects. Fundamental difficulties arose in the case of detecting so-called exo-Earths because of their smaller masses. But the quest for Earth-like planets nevertheless replaced cosmology as the burning question of the 2020s. And the inability of spacecraft to make the journey even to the nearest stars only threw the problem into stark relief.

In an oft-repeated analogy, the Sun and its nearest neighbour star, Proxima Centauri, were likened to two marbles. On that scale, their separation of 4.2 light years shrinks to 300 kilometres—the distance from Sydney to Canberra, for example, or from London to Liverpool, or New York to Boston.

Not very far—but on the same scale, the speed of a chemically powered spacecraft travelling between them is the speed at which grass grows. Little wonder the public lost interest in space research and astronomy.

Such was the decline in enthusiasm for science during the 2030s that talk of an intellectual dark age was common, with a huge majority of the population listing fly-on-the-wall virtual reality TV as their principal spare time activity. The dumbing-down of science culminated in the mass hysteria of September 2040, as the five naked-eye planets aligned within a ten-degree circle in the evening sky. But the discovery the following year of asteroid 2041 FU changed all that.

With the threat of impact eliminated by mid-century and newly developed space technology readily available, attention turned once again to the questions facing astronomers. How could exo-Earths in the Sun's neighbourhood be discovered? And how could astronomers determine whether or not they harboured intelligent life?

In 2043, a conference was held to celebrate the centenary of the birth of a celebrated French astronomer, Antoine Labeyrie. This man had been a professor at the Collège de France in Paris, and had consistently thought outside the square in the design of large telescopes. On the very eve of the twenty-first century, he had proposed the development of a 'hypertelescope' consisting of 150 orbiting spacecraft, each carrying a 3-metre diameter telescope mirror, spread over an area of space 150 kilometres in diameter. Such an instrument would reveal detail on a scale of *micro*-arcseconds—a thousand times better than any existing telescope. With suitable equipment to eliminate the 100 million times greater glare of the parent star, it would be sensitive enough to portray a recognisable image of an Earth-like planet at a distance of 10 light-years.

In fact, a Labeyrie-style hypertelescope called Super-Darwin had been flown by the European Space Agency in the late 2010s, but had failed to detect any exo-Earths within its range of sensitivity. The project had been largely forgotten—but, with the centenary conference and the renewed interest in space and astronomy, it found itself once more on the agenda by the late 2040s.

This time, however, a new hypertelescope was proposed, using much bigger mirrors incorporating another of Labeyrie's brainwaves. That was to make the mirrors not from glass-ceramic, but from delicate reflective membranes, held in perfect shape in zero gravity by the pressure of light itself, emitted from infrared lasers. And this telescope was to have something quite new to optical astronomy—the ability to seek and analyse very short pulses of faint light. That facility had been the province of radio astronomers engaged in the search for extraterrestrial intelligence (SETI) for nearly a century, and no convincing signal had ever been found. But what if ET was using a laser in the visible waveband to advertise that he, she or it had now reached the required level of techno-logical advancement?

With enthusiasm for space research at an all-time high, the new hypertelescope was ready for launch by 2055. Once deployed, early the next year, its one hundred 30 metre gossamer mirrors began to glide silently through space in perpetual orbit around the Moon. The venture did not please everyone, however. When Britain's Astronomer Royal, Sir Pratney Wilbert (2011–2096), was asked about it, he made his feelings clear in language that was intemperate even by the standards of the mid-twenty-first century: 'The membrane-mirror hypertelescope is a load of bollocks, and you'd have to be off your trolley to think it would work.'

But—it did work. In 2058, the first truly Earth-like planet was revealed orbiting the star HD172051 at a distance of 42 light-years from our own Solar System. It showed biomarkers such as the presence of oxygen in its spectrum, and on its surface was a clear analogue of the Amazon Basin on Earth. Despite half a century of watching, however, no hint of signals of artificial origin have been detected coming from the planet in any waveband. Nor from the thirty or so other exo-Earths that the hypertelescope has discovered.

But what the hypertelescope did reveal was an entirely new Universe of exceedingly transient phenomena in the optical waveband. Our Galaxy, for example, is now known to be full of winking lights whose origin and significance are still far from clear. As we celebrate the 500th birthday of the telescope, we look forward with enthusiasm to the latest instrument coming online to answer the new Big Questions. A worthy successor to the creations of Galileo, Herschel, Hale and Labeyrie, it is called the GLT—a one-light-day gravitational lens telescope, formed in space from the first artificial black hole.

NOTES AND SOURCES

PROLOGUE

But human eyes compare (p. xi)
The detectors now universally used
in visible light astronomy are
charge-coupled devices (CCDs),
the light-sensitive microchips that
are also found in video-cameras
(see, for example Leverington 1995,
pp. 308–9).

1. POWER TELESCOPES

The symposium's title (p. 1) 'Power
Telescopes and Instrumentation
into the New Millennium' (27–31
March 2000) was one in a continuing series of international symposia
on large telescopes, co-sponsored by
the European Southern Observatory
(ESO) and SPIE, the International
Society for Optical Engineering.
Like all dimensions in the sky (p. 4)
The resolution (or resolving power)
of a telescope is traditionally
defined as the separation of the
closest pair of stars that it will show
as distinct objects. Modern defini-
tions are more subtle, but the
general principle remains the same.
Putting some figures on resolution
(p. 4) The resolution depends not
only on the diameter of the mirror
or lens, but also on the wavelength
of the radiation with which the
observations are made. As wavelength increases, resolution becomes
poorer. Any given telescope can
record finer detail with visible light
than with infrared radiation, for

example. The figures quoted here
are for green light. Resolution is
also influenced by the quality of the
optical components of a telescope,
and these figures assume a mirror
free from surface inaccuracies and
operating in the absence of atmospheric turbulence. The resolution
obtained under these conditions is
said to be at the 'diffraction-limit'
(because it is limited only by the
fundamental physical phenomenon
of diffraction). See King (1955,
pp. 272–3) for a description of
early work on resolving power,
and any standard text on optics
(for example, Longhurst 1957,
pp. 283–4) for its theoretical basis.

Seeing the unseeable
The Hubble Space Telescope (p. 5)
The telescope operates 600 km
above the Earth's surface and
makes one orbit every 96 minutes.
It is diffraction-limited (see note
above) and detailed technical data
may be found in such publications
as Møller (1996).
That instrument will be (p. 6) The
James Webb Space Telescope
(formerly the Next Generation
Space Telescope) is scheduled for
launch in the early 2010s. Plans for
the instrument and the research it
will carry out are summarised in,
for example, ESA (1998).
A few places emerged (p. 6) An
analysis of the geographical distribution of the world's great optical
telescopes can be found in Watson
(1999b).

By *'instrumentation'* (p. 7) Accounts of the early days of astronomical instrumentation are given in King (1955) and Hearnshaw (1986, 1996).

Power telescopes
Technology had moved on (p. 9) The mirrors of today's 8-metre telescopes fall into three categories:
1 thick, 'conventional' glass mirrors cast in a rotating oven to attain a rough concave shape before optical surfacing (as in the University of Arizona's Large Binocular Telescope and the twin telescopes of the Magellan project);
2 thin, monolithic (i.e., one-piece) meniscus mirrors shaped like a huge watch-glass (as in the two Gemini telescopes, the Japanese Subaru telescope and the four component-telescopes of the Very Large Telescope); or
3 'segmented' mirrors, i.e., mosaics of many smaller mirrors, as in the Keck Telescopes and the Hobby-Eberly Telescope. See Marra (2000) for a note on the originator of segmented mirrors. See also Watson (1999b) and page 328 in this volume.

One of these instruments (p. 10) The telescope containing two 8-metre mirrors (both mounted in a common structure) is the Large Binocular Telescope, while the one made up of four (each of which has its own separate structure) is the ESO Very Large Telescope (see, for example, Schilling (2000) and Chen (2000)).

One might have thought (p. 11) The presentations on auxiliary instruments and adaptive optics systems at the symposium appear in SPIE (2000d) and SPIE (2000c) respectively.

Starving the fever
One had been called the ELT (p. 12) The original ELT, a 25-metre derivative of an unusual present-day 8-metre class instrument (the Hobby-Eberly Telescope) is described in Bash et al. (1997). See also Watson (1999b).
On the other hand (p. 12) By 2004, the GSMT and CELT projects were being lumped together as the TMT (Thirty-Metre Telescope).
The logic went like this (p. 13) A summary of the evolution of optical telescope projects with apertures greater than 25 metres leading to the 100-metre OWL proposal, together with the scientific rationale for such telescopes, is given by Gilmozzi and Dierickx (2000). See also Mountain and Gillett (1998) for a further discussion of these issues. The presentations on extremely large telescopes at the Munich Symposium appear in SPIE (2000a, b).
Could the fact (p. 14) OWL, like the VLT, is a project of the European Southern Observatory.

2. THE EYES OF DENMARK

Two hot-headed young men (p. 18) The description of the duel between Tycho (Tyge) Brahe and Manderup Parsberg is based on details given in Chapter 1 of Thoren (1990), the definitive account of Tycho's life and work. Other episodes of Tycho's early life described in the present chapter are similarly attributed, unless cited otherwise. (Victor E. Thoren was the late twentieth century's foremost authority on Tycho. His career, like Tycho's, ended prematurely. An appreciation can be found in Westfall (1991).)

Thirty-five years later (p. 19) Manderup's protestations of lifelong friendship were made in reply to a published funeral oration for Tycho, which alluded to the duel (Thoren 1990, p. 343).

He was born (p. 19) The Brahe family home at Knutstorp is still used as a residence. The earliest house on the site was built in the mid-fourteenth century, but the present building dates from the reconstruction in 1551. Knutstorp has been the property of the Wachtmeister family since 1771 (Wachtmeister, *c.*1996).

True, the Sun (p. 21) The ancient idea that the markings on the Moon are reflections of geographical features on Earth is discussed by Stooke (1996).

And the planets (p. 21) Mercury, Venus, Mars, Jupiter and Saturn.

Ptolemy's view (p. 21) This was proposed in his great astronomical text, the *Almagest*, during the first century AD. It was a development of much earlier theories, which can be traced back to the time of Plato (4th century BC). Copernicus' Sun-centred theory was published in its final form in his treatise *De revolutionibus* in 1543. Copernicus received a complete printed copy of the work on the day he died.

It reflected (p. 21) A cogent account of the shift in scientific thinking during the Renaissance is given by White (2000), Chapter 2. See also Panek (2000), Chapters 1 and 2.

In the sixteenth century, astrology (p. 23) It was not universally held in high regard, however. Earlier in the

Renaissance, Leonardo da Vinci had described astrology as 'that fallacious judgement by means of which (begging your pardon) a living is made from fools' (quoted in White 2000, p. 47).

Was it still in the air (p. 24) This scenario follows Thoren (1990, pp. 22–3).

Perhaps it was during (p. 24) Thoren (1990, p. 25n) remarks on the early use of skin grafts to correct facial mutilation, suggesting that Tycho would have been able to avail himself of the treatment in the late 1570s had he been aware of it.

Hven

Technically, its name (p. 24) Hven has been part of Sweden since the Peace of Copenhagen in 1660.

The island is small (p. 25) Geographical data and population statistics are from Hydbom (1995).

It served not just as Tycho's home (p. 25) The story of Uraniborg as a research centre, with a 'familia' of collaborators and assistants, is told by Christianson (2000).

This was Stjerneborg (p. 26) Thoren (1990, p. 183), translates the Danish 'Stjerneborg' as 'Star Town', probably because of its rendering in Latin as Stellæburg.

Tycho celebrated (p. 27) The foundation ceremony is described by Thoren (1990, p. 114) and Christianson (2000, p. 53).

But destiny itself (p. 27) A detailed account of the philosophical implications of Tycho's new star and the preparation of *De stella nova* is given in Thoren (1990, Ch. 2). See also Christianson (2000, pp. 17–18).

An audience with his monarch (p. 28) Tycho's own report of Frederick's entreaty to him on 11 February 1576 is quoted in full in Christianson (2000, pp. 22–3).

He built a succession (p. 28) In 1598, Tycho published a description of his most significant instruments in *Astronomiæ instauratæ mechanica* (see Christianson 2000, pp. 223–4), accessible today on the Web at www.kb.dk/elib/lit/dan/brahe/ A particularly clear and complete account of their operation (in Swedish and English) is given by Wennberg (1996); see also Roslund (1989).

In fact, the eye's resolution (p. 29) Under certain circumstances (for example, in pattern-recognition), the eye can perceive detail well below its normal resolution threshold. See, among others, Martin (1948, pp. 159–67) and the historical survey of the perception of detail in Saturn's rings given by Dobbins and Sheehan (2000).

Nevertheless, with cleverly designed (p. 29) A lucid account of atmospheric refraction can be found in Lynch and Livingston (1995, Ch 2).

Frequently, it was as good as (p. 29) Tycho's astronomical achievements—including his steady improvement of positional accuracy—are summarised by West (1997); see also Wesley (1978).

The largest and most spectacular (p. 30) The Great Equatorial Armillary is described and illustrated in Thoren (1990, pp. 174–5) and Wennberg (1996, p. 61). It was not the largest instrument ever built by Tycho; that was a heavy wood and iron quadrant with a radius of 5.4 m, built in Augsburg in 1570. Its accuracy was inferior to Tycho's later instruments (Thoren 1990, pp. 33–4, Roslund 1989).

Legacy

We can judge his certainty (p. 31) Leonardo's diagram appears in the *Codex Leicester*, folio 7r (reproduced in Desmond and Pedretti 2000, p. 75). The hole was noted while the *Codex* was on exhibition in Sydney, 6 September–5 November 2000.

It is to Tycho Brahe (p. 31) Kepler's work, particularly in its relationship to Tycho Brahe, is summarised in Christianson (2000, pp. 299–306).

He was attracted to (p. 32) Tycho's hybrid model of the Solar System is described in his *De mundi ætherei recentioribus phænomenis liber secundus*, published at the end of 1587 (see Christianson 2000, pp. 122–4).

In the 1670s (p. 32) Hevelius justified his adherence to plain sights rather than telescopic sights for positional measurements in his *Machinæ cœlestis* of 1673. His dispute over this matter with the English physicist Robert Hooke is discussed in King (1955, pp. 100–2).

Tycho was not a man to let (p. 33) The foundation of the Uraniborg press (1584) and the paper mill (1592) are described in Christianson (2000, Chs 5 and 6).

Imperceptibly at first (p. 34) Detailed accounts of the events leading to Tycho's departure from Uraniborg can be found in Thoren (1990, Ch 11), and Christianson (2000, Ch 9). These authors also describe Tycho's years in exile and his attempts to found a new Uraniborg (Thoren 1990, Ch 12–13, Christianson 2000, Ch 10).

After a dinner-party (p. 35) Kepler's account of Tycho's death and an eyewitness account of his funeral are quoted in Thoren (1990, pp. 468–70). The modern autopsy findings are outlined in West (1997) and Christianson (2000, p. 413). Both Thoren and Christianson cover Tycho's family life in detail. *Reports issued* (p. 36) King (1955, p. 23) describes the rapid decay of Uraniborg and Stjerneborg (citing J.L.E. Dreyer's *Tycho Brahe* of 1890).

3. ENIGMA

Digges wrote the passage (p. 38) *Pantometria* was published in London in 1571. Its full title is *A Geometrical Practise, named Pantometria, divided into three Bookes, Longimetria, Planimetria and Stereometria, containing Rules manifolde for mensuration of all lines, Superficies and Solides: with sundry strange conclusions both by instrument and without, and also by Perspective glasses, to set forth the true description or exact plan of an whole Region: framed by Leonard Digges Gentleman, lately finished by Thomas Digges, his sonne.* (Its title page is reproduced in Ronan 1991.) The passage quoted appears in a compendium of documents relating to the early history of the telescope assembled and translated by Van Helden (1977b, p. 30) and based on the earlier research of Cornelis de Waard (*De uitvinding der verrekijkers*, The Hague, 1906). Van Helden's analysis of this evidence remains the most scholarly account to date of the origin of the telescope. *From* Pantometria, *and from the writings* (p. 39) Bourne's manuscript text, *A treatise on the properties and qualities of glasses for optical purposes, according to the making, polishing and grinding of them* (*c.*1585), is preserved in the British Library

(MS Landsdowne 121, item 13). It is reproduced in full in Van Helden (1977b, pp. 30–4). The text of Chapter 9 of the document (relating to the suggested telescope) appears in facsimile in Ronan (1991). *Nevertheless, it has led some* (p. 40) Both Ronan (1991) and R.T. Gunther (*Early Science in Oxford*, Oxford, 1921–1945, quoted in Van Helden 1977b, p. 14) interpret Digges' account as a description of a reflecting telescope. See also Mills (1992) and the contributions by Ronan et al. (1993) to a Scientific Instrument Society discussion meeting on the topic. In that discussion, Ronan, G. Satterthwaite and J. Rienitz suggest that Digges used a convex objective lens with a concave mirror as the eyepiece (see Chapter 4). It is true that such use of the mirror would be less demanding on its surface accuracy than if it were used as an objective (see Chapter 8), but it remains unlikely that Digges perfected such an instrument. *It is therefore most unlikely* (p. 40) Bourne's telltale details emerge in Chapter 9 of his *Treatise*, where he mentions mirrors that are 'concave with a foyle, upon the hylly syde'. They therefore resemble a modern make-up or shaving mirror in which the light has to cross an air-to-glass surface, pass through the thickness of the glass and then retrace its steps after reflection from the back surface. King (1955, p. 30) suggests that the backing foil of Digges' mirror was lead. Whatever the material, it would have required an extremely high degree of burnishing. Since Newton's time, telescope mirrors have conventionally been reflective on their front surfaces.

This is a pity, as elsewhere (p. 40) Bourne's description of this rudimentary telescope is quoted from *Inventions or devices. Very necessary for all generalles and captaines, or leaders of men, as well by sea as by land*, published in London in 1578. The text relating to this, the 110th device, is quoted in full by Van Helden (1977b, p. 30).

Remarkably, this is (p. 41) The window telescope is described in Watson (c.1925, pp. 30–1). So-called 'stick telescopes', smaller versions of the same device (for attachment to the end of a walking-stick rather than hanging in a window), were also popular at the turn of the twentieth century.

Bourne's account led (p. 42) King (1955, p. 29) postulates Bourne's hypermetropia. See Martin (1948, pp. 287–8) for changes in the optical characteristics of the eye as it ages.

Clothing the naked eye

While large burning-glasses (p. 43) Such glasses were also described by Bourne. The extraordinarily brief seventh chapter of his *Treatise* consists only of the following sentence: 'The quality of this Glass ys, if that the sunne beams do pearce throughe yt, at a certayne quantity of distance, and that yt will burne any thinge, that ys apte for to take fyer: And this burnynge beame, ys somewhat furder from the glasse, th[a]n the perspective beame'. (That is, the focal point is slightly farther from the lens than the eye-point for viewing distant objects.) Two large, stand-mounted laboratory burning-glasses made a century or so later by Ehrenfried Walther von Tschirnhaus (1651–1708) are described in

Mayr et al. (1990, pp. 68, 81). See also Spargo (1984), and the idiosyncratic account of burning glasses in Temple (2000, Ch 5).

Like the origin of the telescope (p. 43) The steps leading to the introduction of convex and concave lenses in spectacles are summarised by Van Helden (1977b, pp. 10–11) and King (1955, pp. 27–8). A popular account of the history of spectacles—including their much earlier introduction in China—is given by Davidson (1989).

This man wrote (p. 44) The passage from *Magia naturalis* is quoted by Van Helden (1977b, pp. 34–5), in an anonymous translation made in London in 1658. A different translation is used by King (1955, p. 30), but the meaning is identical.

Van Helden interprets (p. 45) Porta's combination of convex and concave lenses is discussed by Van Helden (1977b, pp. 15–19). The use of weak Galilean telescopes as ungainly spectacles to assist with extremely poor sight is noted by, among others, König and Köhler (1959, p. 184).

Bifocal lenses are (p. 46) See, for example, Giscard d'Estaing 1985, p. 221.

Had anyone truly managed (p. 46) Sluiter (1997a, b) charts the remarkably rapid exploitation and spread of the telescope immediately after its appearance in 1608. See also Chapter 4.

His soaring achievements (p. 46) Leonardo's contributions to science are explored in White (2000).

He didn't always arrive (p. 46) White (2000, p. 187), suggests that Leonardo's explanation of the blue of the sky was closer to the truth than it appears. The relevant passage is in the *Codex Leicester*, folio 4r (see Desmond and Pedretti 2000, pp. 50–1). Leonardo realised that the sky was made luminous by the light of the Sun, but was not equipped to understand why. Lord Rayleigh showed in 1871 that it was the scattering of light by the atmosphere. A complete (though elderly) account of the historical development of the theory appears in Humphreys (1920, Ch 7), and a modern description of the phenomenon is given by Lynch and Livingston (1995, Ch 2). *In one of his astronomical observations* (p. 46) An account of Leonardo's observations of the Moon is given by Welther (1999). Leonardo reports his discovery of 'earthshine' in the *Codex Leicester*, folio 2r (see Desmond and Pedretti 2000, pp. 34–5). *But we also have his own* (p. 47) His note to 'construct glasses . . .' is quoted by Gingrich (2000) as well as by White (2000, p. 296). *Indeed, one of the eventual petitioners* (p. 47) The complete text of Metius' patent application is given in Van Helden (1977b, pp. 39–40). See also Chapter 4 of this volume.

Legends and lenses
He had written to the pontiff (p. 48) See the entry on Bacon in the excellent potted biographies of famous mathematicians by O'Connor & Robertson (World-Wide Web). His career is also summarised, for example, by Hall (1995). *Opus maius contains much* (p. 48) This passage is widely quoted (e.g. in King 1955, p. 28; Van Helden 1977b, p. 28; and Ronan 1991). *Bacon is here echoing* (p. 48) Grosseteste's *De iride* was one of a trilogy

of books on light written between about 1220 and 1235 (Ronan 1991). *Other than these words* (p. 49) The reference by Robert Recorde to Bacon's work appears in *The pathway to knowledg, containing the first principles of geometrie* (London, 1551) and is quoted both by King (1955, p. 28) and Van Helden (1977b, p. 29). *It had already been the fate* (p. 49) A short account of Gerbert's career and his impact on contemporary thinking is given by Lacey and Danziger (1999, pp. 188–92). It is to these authors that the 'Bill Gates' epithet is due. *Gerbert built an astronomical clock* (p. 49) This instrument and the use of empty sighting tubes to aid vision is referenced to several sources by Van Helden (1977b, pp. 9–10). These include the relevant quotation from Aristotle's *Historia animalia* (*Generation of Animals*). Sources quoted by Temple (2000, Ch 4), allude to Gerbert's possible use of telescopes. Temple also illustrates a fragment of fifth/fourth-century BC Greek pottery that seems to be depicting the use of a sighting tube (plate 10), but chooses to interpret this as a portrayal of a telescope. *So what of the ancients* (p. 50) The literature referring to the possible knowledge of telescopes by, for example, the Romans, the Carthaginians and the ancient Britons is extensively quoted by Temple (2000, Ch 4). (The passage from Bacon's *Opus maius*, for example, is quoted on p. 128.) Temple concludes from these writings that ancient telescopes were commonplace, but this assertion is unsubstantiated and should be treated with extreme caution.

And earlier still (p. 50) The suggestion that the ancient Assyrians knew of the telescope is made by Giovanni Pettinato (University of Rome) in *La scrittura celeste*. This controversial book received extensive press coverage on its publication (e.g. Bruce Johnston, London *Daily Telegraph*, 1 June 1999). For notes on the territorial ambitions of the Assyrians and the Carthaginians see Kinder and Hilgemann (1974, pp. 29–39).

Is it remotely possible (p. 50) The megalithic sites at Avebury (*c*.1500 BC) and Stonehenge (constructed between *c*.2000 BC and 1400 BC) are described by, among others, Hall (1995). See also Gingerich (1979).

Telescopes had to evolve (p. 51) The true nature of Saturn's rings was first described by Christiaan Huygens in his *Systema saturnium* published in 1659 (King 1955, p. 51).

The most famous one (p. 51) The Nimrud lens and several other ancient lenses are described and pictured in Temple (2000, Ch 1 and appendices).

In fact, glass was quite well known (p. 52) See, for example, Dodsworth (1982).

And a significant hoard (p. 53) A number of Viking lenses are described by Temple (2000, App. 10). Kizer (2000) speculates on their use in telescopes.

Controlling the curvature (p. 54) Van Helden (1977b, p. 16).

There is another (p. 54) Willach (2001).

4. Enlightenment

Ultimately, it was religious dispute (p. 55) Details of the origins and evolution of the Reformation are summarised by, for example, Kinder and Hilgemann (1974), in section 10 ('Age of Religious Discord'). These authors also outline the major events of the Eighty Years' War (1974, p. 245).

The southern provinces (p. 56) Brief accounts of the government and constitution of the United Provinces and the role of Prince Maurice are given by Hoogerdijk et al. (*c*.1994) and Van Helden (1977b, p. 20).

Neither the timing (p. 56) The emergence of the telescope in the context of the 1608 peace conference is noted by Van Helden (1977b, p. 25) and Sluiter (1997b).

The man who turned up (p. 57) The communication from the Ambassador of Siam, dated October 1608, was reprinted as a tract in Lyons the following month (Van Helden 1977b, pp. 40–2; see also Sluiter 1997b, n. 2).

Lipperhey had prepared himself (p. 58) The text of the letter presumed to have been carried by Lipperhey (it does not actually mention the identity of the bearer) is reprinted in Van Helden (1977b, pp. 35–6).

In order to test it (p. 59) An account of the Binnenhof and its history can be found in Hoogerdijk et al. (*c*.1994).

Claim and counterclaim

Whether it was by dint of (p. 60) Prince Maurice's evaluation of Lipperhey's telescope, together with Spinola's reaction to it, is described in the Ambassador of Siam's letter (see note above for p. 57). Sluiter (1997b) disputes the account in respect of Maurice permitting Spinola to see the telescope, but there seems no good reason to doubt it.

On his arrival (p. 60) Sluiter (1997b) explores the possible routes by which news of the telescope might have reached Galileo in Italy, and quotes in full a letter written by Bentivoglio to the papal Secretary of State on 2 April 1609. The letter mentions that Bentivoglio was about to send a telescope to Rome.

The neat, measured handwriting (p. 61) The deliberations of the States General in respect of the telescopes of Lipperhey, Metius and Janssen are recorded in their minutes and correspondence (Van Helden 1977b, pp. 35–43).

It was not until (p. 62) Regarding the origin of the name, most authors quote Edward Rosen (*The Naming of the Telescope*, Schuman, New York, 1947). See, for example, Van Helden (1989, p. 112).

This time it was an instrument-maker (p. 63) Jacob Metius died in 1628. For a note on his brother Adriaen (1571–1635) and Adriaen's relationship with Tycho, see Christianson (2000, p. 322).

Open secret

The later (1614) testimony (p. 65) Marius' account of Fuchs' encounter with the Dutchman is in the preface of his *Mundus jovialis* (1614), which is quoted by Van Helden (1977b, pp. 47–8). Marius lived from 1573 to 1624, and details of his career—including his relationship with Tycho—can be found in Christianson (2000, pp. 319–21).

There is even a hint (p. 65) The identification of Sacharias Janssen as both the young man in Middelburg and the Dutchman in Frankfurt was made by Cornelis de Waard in *De uitvinding der verrekijkers* (1906); see Van Helden (1977b, p. 22). De Waard also uncovered Janssen's exploits on the wrong side of the law.

For what it is worth (p. 65) Van Helden (1977b, p. 23).

We know, for example (p. 66) Willach (2001).

That day's entry (p. 67) The date of Lipperhey's burial is noted by Sluiter (1997b).

The successful construction (p. 68) For the development of binoculars, see Watson (1995, 1999a, 2000).

5. FLOWERING

Galileo Galilei, a professor of mathematics (p. 69) Galileo's family tree is given by Sobel (1999, pp. 14–15).

We know it by its Latin name (p. 69) The translation of *Sidereus nuncius* used throughout Chapter 5 is Van Helden's (Galilei, 1610).

No fool, this Galileo (p. 70) Details of Galileo's tactics leading to the appointment in Florence are given by Van Helden (1989, pp. 9ff).

A French diarist (p. 71) The journals of Pierre de l'Estoile (1546–1611) are quoted by Sluiter (1997b); see also Van Helden (1977b, p. 44).

This man was (p. 71) Harriot's work is discussed briefly by King (1955, pp. 39–40), while his maps of the Moon are reproduced in Bloom (1978).

It is a charitable epithet (p. 72) The extract is from a letter to Harriot (King 1955, p. 40).

Subsequently, this became (p. 73) A brief account of the dispute is given by Van Helden (1977b, p. 27n).

Marius explains (p. 73) Van Helden (1977b, pp. 47–8).

The starry messenger

There are at least two possible routes (p. 74) Sluiter (1997b).

They were arrived at in 1621 (p. 74) A brief account of Snel's career is given by Christianson (2000, pp. 358–61).

In fact, we now know (p. 74) See the entry on Harriot in O'Connor & Robertson (World Wide Web).

A succession of improvements (p. 74) Recent tests by Greco et al. (1992) and Molesini (2003) show that Galileo's telescopes were of extremely high optical quality.

The first two of Johannes Kepler's (p. 76) See Caspar (1959, pp. 123–42) for their development.

Late in 1610 (p. 76) Galileo's observations after the publication of *Siderius nuncius* are described by Van Helden (1989, pp. 102ff.); and King (1955, pp. 37–9).

But such outright Copernicanism (p. 76) Nicholl (2001) gives a brief account of the life of Giordano Bruno (1548–1600). See also Van Helden (1989, p. 97).

A new book (p. 77) This was Galileo's *Letters on Sunspots* (see Koestler 1959, pp. 430–1). For general accounts of Galileo's trial before the Inquisition and his later life, see King (1955), pp. 40–1); Thiel (1958, pp. 142–56); Koestler (1959, Part 5); Ley (1963, Ch 6). A recent extensive analysis of Galileo's trial is given by Westfall (1989).

Tycho's protégé

Unlike the affable (p. 77) For Kepler's character, piety, etc. and distaste for washing see Caspar (1959, pp. 368–9) and Christianson (2000, pp. 299–306).

Kepler's relationship with Tycho (p. 78) For this, and his appoint-ment as Imperial Mathematician, see Caspar (1959, pp. 116–22) and Christianson (2000, pp. 299–306).

But very soon afterwards (p. 78) For his reaction to *Siderius nuncius* see Caspar (1959 pp. 192–8) and Van Helden (1989, pp. 94–9).

Kepler published his results (p. 79) Caspar (1959, pp. 198–202).

Probably the first successful example (p. 82) King (1955, p. 45); Van Helden (1977a).

But bigger events (p. 82) Caspar (1959, pp. 204–8).

Here, he completed (p. 82) Caspar (1959, pp. 264–90).

Further work enabled him (p. 83) For the *Rudolphine Tables* and Kepler's death, see Christianson (2000, p. 305).

6. EVOLUTION

It is a measure (p. 84) Van Helden (1977a).

It was simply that (p. 84) Caspar (1959, p. 201).

On top of that (p. 85) Van Helden (1977a); Willach (2002).

For example, in November 1614 (p. 85) Sluiter (1997a).

A spate of astronomical discoveries (p. 85) See, for example, King (1955, p. 46).

A Capuchin monk (p. 87) For notes on Rheita and Wiesel, see Van Helden (1977a); König and Köhler (1959, pp. 439–40); and Willach (2002).

The seven-year conflict (p. 88) See Hall (1995).

A few years earlier (p. 88) King (1955, pp. 94–7).

Spider threads were perfect (p. 89) On their disuse and rediscovery, see Brooks (1989).

William Gascoigne also invented (p. 89) See Brooks (1991) for an account of early eyepiece micrometers.

The starrie tubus
They obviously got on well (p. 89) Van Helden (1977a).
By now, telescopes had grown (p. 90) For the dimensions of Galileo's 30-times telescope see King (1955, p. 43). Wiesel's pricelist is preserved in the British Museum ('Sloane' 651, 169–71). It is quoted in König and Köhler (1959, p. 440) and Van Helden (1977a).
Sir Charles Cavendish himself (p. 92) Van Helden (1977a).
Eventually, the obsession (p. 92) Simpson (1992).
There is, however, another (p. 92) The effect of chromatic aberration is actually more than 1000 times greater than spherical aberration: Newton (1730, p. 100); see also King (1955, p. 68) and Chapter 9 of this volume.
A letter survives (p. 93) Samuel Hartlib Papers, Bundle 8, iii. The letter is quoted in full in Van Helden (1977a). See also Willach (2002).
While other contemporary opticians (p. 95) The nineteen-lens telescope was built by Eustachio of Naples (King 1955, p. 56). Robert Hooke also used compound lenses (König and Köhler 1959, p. 440).

Innovations
There was great rejoicing (p. 96) See Hall (1995) for notes on the Restoration and the Pepys quotation. See also Tomalin (2002, Ch 7).
In the meantime (p. 96) See Yapp (2000, p. 309).
In London on the day (p. 96) Van Helden (1977a) describes the events

at Reeve's house on the day of Charles' coronation and quotes Huygens' letter to his brother. Simpson (1985) discusses the work of Reeve (whose name is also spelled Reive or Reeves—see King 1955, p. 62).
The trick he had discovered (p. 98) The optical recipe for Huygens' eyepiece is quoted by King (1955, pp. 54–6).
Although Reeve (p. 98) For the work of Cock, Cox et al., see King (1955, p. 62). An early (1673) Cock telescope is illustrated and described in Thoday (1971).
On 4 March 1675 (p. 99) The origins of the Royal Observatory are described, for example, by McCrea (1975, pp. 5–7). For Flamsteed's instruments, see, for example, King (1955, p. 63). The relationship between Abraham Sharp and Flamsteed is described in Harrison (1963), which cites William Cudworth's *Life and Correspondence of Abraham Sharp* (1889).

Dinosaurs
Another was a man (p. 100) For a recent appraisal of Hevelius' work, see Chapman (2002).
In the hands of such men (p. 100) For an account of the development of long focal-length refractors, see King (1955, pp. 49–65).
The Danzig brewer's ambitions (p. 102) Hevelius' proposed tower observatory is illustrated, for example in König and Köhler (1959, p. 441).
Unfortunately, Hevelius' plans (p. 102) See, for example King (1955, p. 54). The letter to Louis XIV is quoted by Thiel (1958, pp. 157–8).

Three of Huygens' long-focus objectives (p. 105) See King (1955, pp. 63–4), Thoday (1971). The poor quality of glass in Huygens' lenses is noted by Turner in Ronan et al. (1993, p. 5). See also Molesini (2003) for a recent scientific assessment of seventeenth-century optical glass.

7. ON REFLECTION

They have paid little more (p. 107) For an introduction to Islamic culture, see Bloom and Blair (2000).
Of particular relevance to our story (p. 107) Al-Haytham's full name is given in O'Connor and Robertson (World Wide Web). See also Bloom and Blair (2000).
Alhazen's work was championed (p. 107) See King (1955, p. 26); Ronan (1991).
Astonishingly, both Vitello (p. 108) The quotation is from 'The Squire's Tale'; see Chaucer (c.1387), Fragment V (Group F) and Ronan (1991). Chaucer's scientific interests are mentioned by Ronan (1991) and Wright (1985).
Letters between Galileo (p. 109) See Danjon and Couder (1935, p. 605) and Ariotti (1975).
In fact, as it turned out (p. 109) Zucchi's experiment is described by Danjon and Couder (1935, p. 608) and King (1955, p. 44).
What Zucchi didn't realise in 1616 (p. 110) Watson (2002) compares the surface accuracy needed for lenses and mirrors, and describes the bathtub experiment.
We know that even by the 1640s (p. 112) See Willach (2001).

Telescopes of the imagination
And it led to one of the most (p. 112)

Wilson (1996, p. 10) identifies this paradox.
Less well-known is (p. 113) See the entry on Descartes in O'Connor and Robertson (World Wide Web).
It is Descartes' work on optics (p. 114) Descartes and Mersenne's contributions are described by King (1955, p. 48) and Wilson (1996, pp. 2–6).
Sadly, rather than praising (p. 117) See Danjon and Couder (1935, p. 609); King (1955, p. 48); and Wilson (1996, p. 5).

8. MIRROR IMAGE

St Andrews Cathedral was (p. 118) See Lamont-Brown (1989, Ch 3).
When I was a student (p. 119) O'Connor and Robertson (World Wide Web) include entries on both Copson and Gregory.
Nevertheless, among mathematicians (p. 120) See entry on Gregory in O'Connor and Robertson (World Wide Web); Simpson (1992); and Cant (1970, p. 74).
Gregory was not without resourcefulness (p. 120) O'Connor and Robertson (World Wide Web) and Cant (1970, p. 75).
What is it about Gregory's work (p. 121) Simpson (1992) describes the telescope and Reeve's attempt to make the optics.
But the polishing (p. 123) Letter from Gregory to John Collins (1625–1683), a prominent Fellow of the Royal Society, dated 7 March 1673 (Turnbull 1959, pp. 258–61).
This Gregory declined (p. 123) Letter from Gregory to Collins, 23 September 1672 (Turnbull 1959, pp. 239–41).

In fact, it was Robert Hooke (p. 124) King (1955, p. 77).

Genius and skill
Richard Reeve, besides being (p. 124) An account of Reeve's work, trial for manslaughter and death is given by Simpson (1985). Modern optical tests on surviving examples of Campani's work confirm the perfection of his telescopes: see Molesini (2003). For descriptions of the Great Plague, see Hall (1995) and Tomalin (2002, Ch 11).
Born on Christmas Day (p. 126) Newton's life and work are summarised in his entry in O'Connor and Robertson (World Wide Web).
That was the year (p. 127) For Newton's experiments on colour see Newton (1730), Book One; King (1955, pp. 68–71); and Thiel (1958, pp. 173–8).
Convinced of the impossibility (p. 127) Newton (1730, p. 102).
A throwaway line (p. 127) Newton (1730, p. 106).
He made an alloy (p. 127) King (1955, p. 74).
Next, Newton had to figure out (p. 128) Newton (1730, p. 104).
Newton's first specimen (p. 129) He describes his telescope and its performance compared with a refractor in Newton (1730, p. 103). Note that the focal length of a spherically curved mirror is half its radius of curvature, which is why Newton quotes the length of his telescope as a quarter of the 'diameter of the Sphere'.
But he made a second example (p. 130) King (1955, p. 74). The provenance of the Royal Society's model of Newton's telescope is discussed by Bishop (1980).

The theory completed
A scientist called de Bercé (p. 131) Danjon and Couder (1935, p. 613); King (1955, p. 75); Turnbull (1959 p. 151n); Wilson (1996, p. 9); and Baranne and Launay (1997).
For more than 300 years (p. 131) See, for example, Bell (1922, p. 22); King (1955, p. 75); and Wilson (1996, p. 470).
Then, in a remarkable piece (p. 131) Baranne and Launay (1997).
What is more surprising (p. 132) Simpson (1992). Gregory reveals his attempt in a letter to Collins dated 23 September 1672: 'yet I made some tryals, both with a litle [sic] concave & convex speculum' (Turnbull 1959, pp. 239–41). As it eventually turned out, a working Gregorian was easier to make than a working Cassegrain because of the difficulty in optically testing a convex secondary mirror. It is to Jesse Ramsden (see Chapter 9) that the first serious attempt at a Cassegrain is usually attributed (Wilson 1996, p. 15).
It was exactly this point (p. 133) Letter to Oldenburg of 4 May 1672 (Turnbull 1959, pp. 153–5).
Up in St Andrews (p. 133) Simpson (1992). Gregory responded in letters to John Collins (who was also in correspondence with Newton) dated 23 September 1672 (Turnbull 1959, pp. 239–41) and 7 March 1673 (Turnbull 1959 pp. 258–61).
No such lengthy twilight years (p. 134) Gregory's dissatisfaction at St Andrews, move to Edinburgh and early death are described by O'Connor and Robertson (World Wide Web).
But there exists an astonishing book (p. 135) Ariotti (1975).

9. SCANDAL

It is hard for us (p. 137) Waller (2000, pp. 309–11).

The well-heeled might seek lodging (p. 137) The highwayman was John Hall, quoted in Waller (2000, p. 376). Executions at Tyburn are described by Waller, (2000, pp. 327–32).

He had also made a startling discovery (p. 138) Newton (1730, p. 100).

But—erroneously—he also put paid (p. 139) Newton (1730, p. 102).

It was a nephew of (p. 139) See O'Connor and Robertson (World Wide Web) for notes on David Gregory and his work.

He suggests that (p. 139) The quotation is from the second (1735) edition of *Catoptricae et dioptricae sphericae elementa*: King (1955, p. 144).

In fact, this idea (p. 140) Newton (1730, pp. 101–2).

The idea of combining lenses (p. 140) Barty-King (1986), p. 21.

Indeed, he was a rather accomplished lawyer (p. 140) Chester Moor Hall's life, profession, invention of the achromatic lens and death are described by Barty-King (1986, pp. 22, 46).

In fact, Newton himself (p. 141) King (1955, pp. 68–71).

Hall selected for his experiments (p. 141) Descriptions of crown and flint glass can be found in Dodsworth (1982, pp. 9–10).

Scarlett and Mann were among (p. 142) See Simpson (1985) for notes on Scarlett, and Talbot (2002) for Mann.

Success and obscurity

This man was Jesse Ramsden (p. 143) For Ramsden's life and

work, see King (1955, pp. 162–72).

In his Royal Society document (p. 144) The full text is given by Talbot (1996).

Recent circumstantial evidence (p. 145) See Talbot (2002).

Ramsden also speaks of a telescope (p. 145) Talbot (1996).

It was another cloth-worker (p. 146) For Dollond's early life, the origins of Dollond & Son and Dollond's correspondence, see Barty-King (1986, Ch. 1) and King (1955, pp. 145–50).

During that same voyage (p. 147) Haynes et al. (1996, Ch 2).

Another sympathetic ear (p. 147) For Short's life and work, see Clarke et al. (1989, pp. 1–10).

No doubt it was with his head spinning (p. 148) For Dollond's encounter with Bass, and the development of his achromatic lens, see Barty-King (1986, Ch 1); King (1955, pp. 145–8); and Talbot (1996).

Unbridled bitterness

But that was when the trouble started (p. 148) For the Dollond patent and its success, John's death and the court cases, see Barty-King (1986, Ch 1); King (1955, pp. 154–5); and Talbot (1996).

There might be some (p. 153) For Ramsden's marriage and his later dispute with Peter, see Barty-King (1986, pp. 42–3) (note that the year quoted for the marriage is incorrect); King (1955, p. 162); and Talbot (1996).

Such lenses can withstand (p. 154) For triplet lenses and Peter Dollond's later products, see King (1955, pp. 156–60).

The Dollond company went (p. 154) For Dollond telescopes at war, and

Peter's retirement and death see Barty-King (1986, pp. 79–82).

10. THE WAY TO HEAVEN

When Isaac Newton threw down (p. 156) See Chapter 8 and references therein.

Suddenly, here was a telescope (p. 157) For a description of Hadley's work, the Hadley telescope and its performance compared with the 123 ft see King (1955, pp. 77–84); and Bell (1922, pp. 24–7).

The surface must be paraboloidal (p. 159) Brooks (2001a, b) gives more details of 18th-century mirror-making.

Like his friend John Dollond (p. 160) James Short's early life, the development of his career and his telescopes are described in Clarke et al. (1989, pp. 1–5); see also Bryden (1972).

By the time of Short's death (p. 163) His largest telescopes, and King's comment on his work can be found in King (1955, p. 85). The 18-inch aperture telescope dates from 1742 and survives today in the Museum of the History of Science, Oxford (Chapman 1998, p. 342, n. 15).

Heavenly musician

'If not the greatest astronomer' (p. 163) This comment on Herschel's work is paraphrased from Wilson (1996, p. 15).

He was already an accomplished musician (p. 164) For an account of Herschel's early life see Hirshfeld (2001, Ch 10).

Among the latter are (p. 164) The recording of Herschel's organ works is *Pièces d'orgue de William Herschel* by Dominique Proust issued by Disques Dom, Vincennes (DOM CD 1418); Herschel's musical works are summarised on the sleeve-note: Proust (1992).

They were traditional spindly refractors (p. 165) King (1955, pp. 120–4). Herschel's comment about the refractors is from J.L.E. Dreyer's *Scientific Papers of Sir William Herschel* (1912), quoted by King.

Soon, it was followed by an exquisite (p. 167) Details of Herschel's 7-, 10- and 20-ft telescopes are given by Bennett (1976).

Simply the best

A little more than two years later (p. 168) See, for example, Hirshfeld (2001, pp. 180–1). Confirmation by Lexell and naming of the planet: Hoyle (1962, p. 164).

But Herschel's appointment (p. 169) The royal pension, Watson's part in it, the 30-ft and Large Twenty-foot telescopes: see King (1955, pp. 124–7).

Fortunately, his telescopes (p. 169) The telescope-making business earned him at least £15 000 and possibly as much as £20 000 (Spaight 2004).

Herschel's younger brother (p. 170) Bennett (1976).

Herschel's main scientific projects (p. 172) King (1955, pp. 127–8); Bennett (1976).

His aspirations contrast (p. 173) For the conventional positional work of eighteenth century observatories, see, for example, Bennett (1992) and Turner (2002).

Quantum leap

In an early forerunner (p. 174) Bennett (1976); King (1955, pp. 128–9); Hoskin (2003).

These were fairly generous amounts (p. 174) Caroline's comment is quoted by Turner (1977).

In the meantime (p. 175) For the construction and fate of the Forty-foot, see King (1955, pp. 129–34); Bennett (1976).

Procedure (p. 175) Bennett (1976); King (1955, p. 128).

Equipment (p. 175) King (1955, p. 139); Hirshfeld (2001, p. 178); Bennett (1976).

Observing conditions (p. 176) Bennett (1976).

Health and safety (p. 176) King (1955, p. 128); Hirshfeld (2001, p. 182); Hoskin (2003).

By the turn of the new century (p. 177) The last observation with the Forty-foot is noted by King (1955, p. 133), but see the correction in Hoskin (2003).

Despite the lack of regular usage (p. 177) Hoskin (2003).

Later in his life (p. 177) Bennett (1976).

Polishing (p. 178) Hoskin (2003).

He also left behind (p. 179) King (1955, p. 142).

11. ASTRONOMERS BEHAVING BADLY

Take, for example (p. 181) For Andrew Barclay's life and work, see Clarke et al. (1989, pp. 197–202).

As late as the 1820s (p. 184) Barty-King (1986, p. 92).

Part of the problem (p. 184) King (1955, p. 189).

This 'window tax' (p. 184) Hall (1995).

The bigger problem (p. 184) King (1955, p. 176) and Hirshfeld (2001, p. 232).

So acutely was (p. 184) King (1955, p. 189) and Barty-King (1986, p. 95).

During the late 1790s (p. 184) For Guinand's work and move to Benediktbeuern see: King (1955,

pp. 176–9). Details of his techniques are given in Riekher (1990, Ch 7).

Within a few years (p. 185) The most prolific German instrument maker of the British monopoly period was Georg Friedrich Brander (1713–1783); see Brachner (1983).

Whiz kid

This unfortunate lad (p. 185) For Fraunhofer's early life and recruitment see Hirschfeld (2001, Ch 13); King (1955, pp. 178–9).

But his arrival in (p. 186) For Fraunhofer's work and the Great Dorpat Refractor see: Hirschfeld (2001, Chs 13 and 14); King (1955, pp. 180–8); and Riekher (1990, Ch 7).

This instrument was made (p. 189) See, for example, Brooks (1991).

His boyhood frailty (p. 190) Hirschfeld (2001, pp. 242–3); King (1955, p. 188).

All-out war

When Pierre Guinand (p. 190) For his later work, and glass sales to Cauchoix, see King (1955, pp. 179–80).

The combatants in (p. 191) For James South, Richard Sheepshanks and their feud, see Hoskin (1989, 1991); McConnell (1992, pp. 29–30).

When the completed telescope (p. 193) McConnell (1994).

This was a fatal flaw (p. 194) Chapman (1998, p. 43).

They take the form (p. 194) Simms' exercise book see McConnell (1994).

Early twentieth-century historians (p. 196) Dreyer and Turner (1923, pp. 52–5).

It languished (p. 197) The occasional use of South's lens on a temporary stand is noted by Dreyer and Turner (1923, p. 52n), and Duncan Steel in Hoskin (1991). South gave the lens to Dublin University on the occasion of Lord Rosse's installation as Chancellor, 17 February 1863 (Glass 1997, pp. 29–32). The gift became the objective lens of the South Telescope at Dunsink Observatory, built by Thomas Grubb (see Chapter 12). The instrument is still in use.

12. LEVIATHANS

The glue that held together (p. 198) For Romney Robinson and Thomas Grubb, see Glass (1997, pp. 9–11).

He soon had the opportunity (p. 199) Grubb's Markree telescope is described in Glass (1997, pp. 13–16); and McKenna-Lawlor and Hoskin (1984).

This spectacular engineering (p. 200) Chapman (1998, p. 49), has a more favourable assessment of the telescope's productivity than Glass (1997, p. 15).

Hot on the heels (p. 200) The Armagh reflector and its equilibrated lever support system are described in Glass (1997, pp. 17–19). Such a mirror support system is today often known as a whiffle-tree, after the American term for the crossbar that equalises the pull in a horse's harness.

Happily, the pioneering (p. 201) I was privileged to see the fully restored telescope being crated in David Sinden's workshop for its return to Ireland, February 2004.

The success of this (p. 201) Glass (1997, p. 21).

His fortune had already been (p. 202) Birr Scientific and Heritage Foundation (World Wide Web).

The future Lord Rosse (p. 202) For his early telescopes, see King (1955, pp. 206–9) and Hoskin (2002).

Significantly, he adopted (p. 202) Quotation from *Phil. trans.* (vol. 130, pp. 503–27, 1840); Glass (1997, p. 22). See also Wilson (1996, p. 404).

The segmented 36 inch mirror (p. 203) For the intended use of Rosse's telescopes, see: Hoskin (2002).

The tests established (p. 203) King (1955, p. 208); Hoskin (2002).

The historian Michael Hoskin (p. 204) Hoskin (1989, 2002).

Spiral structure

His first attempt (p. 204) Casting and cooling of the 6 ft mirror: King (1955, pp. 210-11); Hoskin (2002).

Lord Rosse adopted (p. 205) King (1955, pp. 212–13).

By February 1845 (p. 207) Early observations and the discovery of spiral nebulae: Hoskin (2002).

The Leviathan of Parsonstown was (p. 208) Dewhirst and Hoskin (1991); Hewitt-White (2003).

Comfort and joy

Like Andrew Barclay (p. 209) For Nasmyth's life and work see Wailes (1963).

He was a noted raconteur (p. 209) For Nasmyth's good humour and relationship with Flossie (Russell?) see Chapman (1998, pp. 108, 350 n. 99).

His name lives on (p. 210) King (1955, p. 217); Wailes (1963); and Thoday (1971).

This arrangement (p. 211) Nasmyth's description can be found in Chapman (1998, p. 348 n. 82).

Lassell eschewed the complexity (p. 213) Wilson (1996, p. 473).

It was not that much harder (p. 213) King (1955, p. 218).
Less than two weeks (p. 213) Moore (1996, p. 39). The 150th anniversary of the discovery of Triton was commemorated in 1996 by the inauguration of a full-scale replica of the 24-inch telescope in Liverpool. It uses one of the two original mirrors (Chapman 1998, pp. 110, 125).
The innovation embodied (p. 214) King (1955, pp. 220–4); see Chapman (1998, p. 107) for Nasmyth's suggestion of the open tube; also Wilson (1996, pp. 406–8). See Wilson (1999, pp. 257–8) regarding Lassell's astatic mirror support.

13. Heartbreaker

Astronomers were painfully aware (p. 216) King (1955, pp. 200–3) and Bennett (1976).
His tone was (p. 217) Quoted in Warner (1982).
It was intercepted (p. 217) Warner (1982) and Glass (1997, pp. 39–40).
A couple of years later (p. 217) Warner (1982); Glass (1997, p. 40); and Haynes et al. (1996, p. 98).
Its radical design (p. 218) Glass (1997, pp. 42–3).
Buoyed by this (p. 218) Warner (1982) and Glass (1997, p. 44).
That might have been the end (p. 218) For Victoria's prosperity and Wilson's appointment, see Haynes et al. (1996, pp. 98–9); Glass (1997, p. 44); Gascoigne (1996).
What is more remarkable (p. 219) For Verdon's role and the Royal Society's response, see Haynes et al. (1996, p. 99).

Engineering masterpiece
Despite these clear advantages (p. 220) Haynes et al. (1996, p. 101); Glass (1997, pp. 46–7); and Gascoigne (1996).
Further controversy came (p. 220) Glass (1997, pp. 44, 49) and Gascoigne (1996).
Three decades later (p. 221) Glass (1997, p. 49). For the changing name of Grubb's firm over the years, see Burnett and Morrison-Low (1989, p. 125) and Anderson et al. (1990, p. 35).
Grubb followed a similar (p. 222) For a description of the manufacture and testing of the telescope, see Glass (1997, pp. 49–58).
At last, the great instrument (p. 223) Haynes et al. (1996, p. 103).

Decline and disaster
It soon emerged (p. 223) The poor image quality, Le Sueur's repolishing and resignation, the problems with the mirrors and Gascoigne's conclusion are all described in Gascoigne (1996).
Whatever the real reason (p. 226) King (1955, p. 267).
After the closure of (p. 226) For developments after the Second World War, see Haynes et al. (1996, pp. 111–13); Gascoigne (1996); Hart et al. (1996); and Frame and Faulkner (2003, Chs 6, 7 and 10).
But then disaster struck (p. 227) Frame and Faulkner (2003, pp. 271–6).

14. Dream Optics

An up-and-coming (p. 230) For Brahms and the *German Requiem* see Holmes (1987, Ch 7).

Like Newton (p. 231) Newton to Hooke, 5 February 1676 (Turnbull 1959, p. 416).

Both are written in (p. 231) Highly readable introductions to notation and structure in music and mathematics are provided respectively by Károlyi (1965) and Acheson (2002).

This man was Ernst Abbe (p. 231) An appraisal of Abbe's achievements (from the perspective of the DDR) can be found in Schütz (1966).

In the hands of these men (p. 231) Wilson (1996, pp. 472–6).

During the 1870s (p. 232) Martin (1948, Ch 8).

In January 1881, he met (p. 232) See, for example, Zeiss (1996, pp. 6–12).

New types of (p. 233) See Watson (1995, 1999a) for the development of Zeiss binoculars.

And then, during the first (p. 233) King (1955, pp. 346–50).

Sifting starlight

Other large instruments (p. 234) Riekher (1990, pp. 193–7); King (1955, pp. 248–9).

The refracting telescope (p. 234) An account of the life and work of Thomas Cooke is given by McConnell (1992, Chs 6–8).

It had been ordered (p. 234) Huggins and his refractor are described in Chapman (1998, pp. 114–15).

Isaac Newton was only (p. 236) Newton to Oldenburg, 6 February 1672 (Turnbull 1959, p. 92).

Newton's discovery (p. 236) For an account of the early work in spectroscopy, see Hearnshaw (1986, Chs 2–4). A general introduction is given by Thackeray (1961).

Then, in 1864, along came (p. 239) Hoskin (1982, pp. 151–2).

While it took another sixty years

(p. 240) See, for example, Watson (2003a) for an account of the details and how they were sorted out.

Record breakers

In 1870, Thomas Cooke (p. 241) For details of the Newall refractor, see McConnell (1992, p. 57) and King (1955, pp. 252–4).

The Newall Telescope's moment (p. 241) Clark and the Washington refractor are described in King (1955, pp. 255–9).

This time, the winning (p. 241) A description of Grubb's Vienna refractor is in King (1955, p. 306) and Glass (1997, Ch 4).

During the second half (p. 242) For Grubb and the Lick Telescope, see Glass (1997), Chapter 5.

James Lick was a (p. 242) Misch and Stone (World Wide Web).

Only two refractors (p. 244) There was to be a third refractor greater than 1 metre in aperture, but it was never completed. It was intended for the Pulkovo Observatory in Russia. The telescope and dome were completed in 1929, but the glass blanks for the 41-inch (1.04 m) objective were not accepted by Soviet astronomers, and the project stalled (see Grubb Parsons 1926, Figs. 2, 6; Warner 1975).

The first was the Yerkes (p. 244) King (1955, pp. 314–18).

This ill-fated instrument (p. 245) The Great Paris Exhibition Telescope is described by Brenni (1996), and Débarbat and Launay (2002), who also report the rediscovery of the lenses.

As diplomatic tensions grew (p. 247) For descriptions of the military optical instruments of the era see Gleichen (1918); König and Köhler

(1959); and Moss and Russell (1988, Chs 1–3).

It is a measure of (p. 247) For optical munitions and the attempted British purchase of German optical instruments, see Reid (2001, Ch 1); Reid (1983); and McConnell (1992, pp. 75–6).

15. SILVER AND GLASS

'Le telescope de Lord Rosse' (p. 248) Gascoigne (1996).

A year earlier (p. 249) For descriptions of the Foucault and Steinheil telescopes, see King (1955, p. 262); Glass (1997, p. 46); and Riekher (1990, p. 223).

In fact, the new technology (p. 249) For a comparison of glass and metal mirrors, see Gascoigne (1996); and Riekher (1990, p. 224).

It was not until 1872 (p. 249) Glass (1997, pp. 69–70). For Piazzi Smyth, see Brück (1983, Chs 4–5).

This 'knife-edge' test (p. 249) King (1955, pp. 262–4); and Riekher (1990, pp. 224–7).

Following pioneering studies (p. 250) The definitive mid-twentieth-century account of astronomical photography and photographic telescopes is in Dimitroff and Baker (1945, Chs 3 and 4).

This ambitious project (p. 251) See, for example, Hearnshaw (1996, pp. 136–42). A delightful description of the 13-inch (33 cm) photographic telescope built at Sydney Observatory for the *Carte du Ciel* is given in Russell (1892). Its objective was made by Sir Howard Grubb.

Absolutely nebulous

In 1901, Ritchey was (p. 252) King (1955, p. 327); Riekher (1990,

pp. 267–9); and Wilson (1996, p. 416).

Remarkably, this modest instrument (p. 252) For the comparison with the 40-inch refractor see Leverington (1995, p. 264).

But Ritchey was aware of (p. 252) See Leverington (1995, pp. 284–5).

When Hale procured (p. 253) The 60-inch telescope is described by King (1955, p. 328–32); Riekher (1990, pp. 269–76); and Wilson (1996, pp. 416–19) (where the 'death-knell' quotation appears).

By 1917, their attentiveness (p. 253) Descriptions of the Hooker Telescope are given by King (1955, pp. 332–8); Riekher (1990, pp. 276–81); and Wilson (1996, pp. 419–22). An account of the funding of the Hooker Telescope is given by Hale (1928).

Using a series of photographs (p. 255) The breakthrough in measuring the distances to spiral nebulae was announced in Hubble (1925).

But there was an immediate (p. 255) An account of the feud between Hubble and van Maanen (including the text of Adams' confidential memorandum on the issue) is given by Hetherington and Brashear (1992).

Wider perspective

It was another eccentric (p. 258) Details of Schmidt's life and work are given in the authoritative account by his nephew (Schmidt 1996). It dispels some of the popular misconceptions repeated by Hodges (1953) and other authors.

He was right (p. 258) See Osterbrock (1994) for details of the wartime use of Schmidt-type optics.

What was this wonderful invention (p. 260) Schmidt's one and only scientific publication is a description of his optical system (Schmidt 1931). *Thanks to Baade's enthusiasm* (p. 261) An account of the use of large Schmidt telescopes up to the present time is given by Watson (2001). *Back in 1982* (p. 263) The advantages of Schmidt telescopes for the technique of multiple-object spectroscopy are summarised by Dawe and Watson (1984), while the evolution of the technique is reviewed by Watson (2003b).

Palomar and beyond

We catch a glimpse (p. 264) Hale (1928).

Although its construction (p. 264) The Hale Telescope is described in detail in King (1955, pp. 401–15); Riekher (1990, Ch 16); and Wilson (1996, pp. 427–30).

Its mirror, for example, was made (p. 265) Di Cicco (1986) gives an account of the manufacture and delivery of the Hale mirror blank.

By contrast, the instrument (p. 266) Descriptions of the BTA appear in Riekher (1990, Ch 22); and Wilson (1996, pp. 430–3).

It set the pattern (p. 267) See Chapter 1 in this volume and references cited. Many of the great reflectors of the late twentieth century are illustrated in Moore (1997), but note that some of the captions are incorrect.

On the factory floor

Perhaps more important (p. 267) Descriptions of the 4-metre class telescopes are given in Riekher (1990, Ch 21) and Wilson pp. 433–42).

Some have criticised this (p. 268) A discussion of the effect of the Hale Telescope on subsequent telescope design can be found in Learner (1986).

But without exception (p. 268) A detailed comparison of materials for telescope mirror blanks appears in Wilson (1999, pp. 216–31).

Most used a cleverly contrived (p. 268) The casting of two large Cervit blanks is described in Anon (1969), while a full account of the procurement of the Cervit blank for the Anglo-Australian Telescope is given by Gascoigne et al. (1990, Ch 6).

By the early 1920s, however (p. 268) The end of the Grubb company and its takeover by Parsons is summarised in Glass (1997, pp. 213–25).

Rosse's youngest son (p. 269) Details of Sir Charles Parsons' work on the steam turbine are given in Strandh (1979, pp. 132–5).

Accordingly, in April 1925 (p. 269) *Nature*, vol. 115, p. 581, quoted in Glass (1997, p. 225).

The new firm (p. 269) A summary of Grubb Parsons' astronomical products up to the mid-1950s is given in Grubb Parsons (1956, p. 28).

These two gentlemen (p. 269) Sisson (1989).

While David Brown was (p. 270) The work in computer-assisted optical polishing is also noted by Wilson (1999, pp. 3–4).

Australian astronomer Ben Gascoigne (p. 271) See Gascoigne et al. (1990, p. 97).

16. WALKING WITH GALAXIES

Hot on the heels (p. 272) The velocity–distance relationship in galaxies was formulated by Hubble (1929). See also Leverington (1995, pp. 236–7, Ch 12).

Mounting evidence suggested (p. 273) A cogent summary of the observational evidence for dark matter and dark energy is given by Nicolson (2001).
He noted that (p. 274) King (1955, p.140).
When, in 1932, he discovered (p. 274) The development of radio telescopes is summarised by Leverington (1995, Ch 15). See also Haynes et al. (1996, Chs 8ff.) for the evolution of radio astronomy in Australia.
In Australia, for example (p. 275) See Frame and Faulkner (2003, p. 108).

Telescopes in space
On 1 January 1956 (p. 276) McCrea (1975, pp. 51–66).
On his arrival in Britain (p. 277) Frame and Faulkner (2003, p. 108).
Sometimes, there were unexpected (p. 278) For the story of gamma-ray bursters, see Murdin (1998).
And, as if these invisible radiations (p. 279) An introduction to the astronomy of sub-atomic particles is given by Clay and Dawson (1997).
More useful to survival (p. 279) For the eye of the scallop and Schmidt-type optics see Mills (1993).
During the 1960s, a new class (p. 280) For the discovery of quasars, see Leverington (1995, pp. 237–40).

In 1979, astronomers using (p. 280) Early examples of gravitational lensing are given by Leverington (1995, pp. 241–3).
It does this because gravity (p. 281) The processes involved in gravitational lensing and Einstein rings are described by Natarajan (1998).

EPILOGUE—21.09.2108

The discovery of the Earth-crossing (p. 283) Asteroid 2041 FU is fictitious, but the potential effects of an asteroid impact are not. See, for example, Steel (2002).
It was not so much the technology (p. 284) This section is a tongue-in-cheek extrapolation of current plans, suggesting the kind of developments that might take place.
The dumbing-down of science (p. 287) The alignment of the planets on 8 September 2040 is a real event.
On the very eve (p. 287) Labeyrie (1999).
That was to make the mirrors (p. 288) Labeyrie (1979).
In 2058, the first (p. 289) HD172051 is one of the candidate stars for today's proposals to search for Earth-like planets beyond the Solar System.

REFERENCES

Acheson, David, 2002, *1089 and All That: A Journey into Mathematics*, Oxford.

Anderson, R.G.W., Burnett, J. and Gee, B., 1990, *Handlist of Scientific Instrument-Makers' Trade Catalogues, 1600–1914*, National Museums of Scotland, Edinburgh.

Anon, 1969, 'Giant mirror blanks poured for Chile and Australia', *Sky & Telescope*, vol. 38, pp. 140–3.

Ariotti, Piero E., 1975, 'Bonaventura Cavalieri, Marin Mersenne, and the reflecting telescope', *Isis*, vol. 66, pp. 303–21.

Baranne, André and Launay, Françoise, 1997, 'Cassegrain: un célèbre inconnu de l'astronomie instrumentale', *Journal of Optics*, vol. 28, pp. 158–72.

Barty-King, Hugh, 1986, *Eyes Right: The Story of Dollond & Aitchison Opticians, 1750–1985*, Quiller Press, London.

Bash, Frank N., Sebring, Thomas A., Ray, Frank B. and Ramsey, Lawrence W., 1997, 'The extremely large telescope: a twenty-five meter aperture for the twenty-first century' in Arne Ardeberg (ed.), *Optical Telescopes of Today and Tomorrow: Following in the Direction of Tycho Brahe*, Proc. *SPIE*, vol. 2841, pp. 576–84.

Bell, Louis, 1922, *The Telescope*, McGraw-Hill, New York.

Bennett, J.A., 1976, '"On the power of penetrating into space": the telescopes of William Herschel', *Journal for the History of Astronomy*, vol. 7, pp. 75–108.

——1992, 'The English quadrant in Europe: instruments and the growth of consensus in practical astronomy', *Journal for the History of Astronomy*, vol. 23, pp. 1–14.

Birr Scientific and Heritage Foundation, *Birr Castle Demesne*, http://www.birrcastleireland.com, March 2004.

Bishop, Roy L., 1980, 'Newton's telescope revealed', *Sky & Telescope*, vol. 59, p. 207.

Bloom, Jonathan and Blair, Sheila, 2000, *Islam: A Thousand Years of Faith and Power*, TV Books, New York.

Bloom, Terrie F., 1978, 'Borrowed perceptions: Harriot's maps of the Moon', *Journal for the History of Astronomy*, vol. 9, pp. 117–22.

Brachner, Alto, et al., 1983, *G.F. Brander, 1713–1783: Wissenschaftliche Instrumente aus seiner Werkstatt*, Deutsches Museum, München.

Brenni, Paolo, 1996, 'Nineteenth-century French scientific instrument makers, XI: the Brunners and Paul Gautier', *Bulletin of the Scientific Instrument Society*, no. 49, pp. 3–8.

Brooks, Randall C., 1989. 'Methods of fabrication of fiducial lines for 17th–19th century micrometers', *Bulletin of the Scientific Instrument Society*, no. 23, pp. 11–14.

——1991, 'The development of micrometers in the seventeenth, eighteenth and nineteenth centuries', *Journal for the History of Astronomy*, vol. 22, pp. 127–73.

——2001a, 'Techniques of eighteenth century telescope makers—Part 1', *Bulletin of the Scientific Instrument Society*, no. 69, pp. 27–30.

——2001b, 'Techniques of eighteenth century telescope makers—Part 2', *Bulletin of the Scientific Instrument Society*, no. 70, pp. 6–9.

Brück, Hermann A., 1983, *The Story of Astronomy in Edinburgh from its Beginnings until 1975*, Edinburgh University Press.

Bryden, D.J, 1972, *Scottish Scientific Instrument-Makers, 1600–1900*, Royal Scottish Museum Information Series, Edinburgh.

Burnett, J.E. and Morrison-Low, A.D., 1989, *'Vulgar and Mechanick': The Scientific Instrument Trade in Ireland, 1650–1921*, National Museums of Scotland, Edinburgh, and The Royal Dublin Society, Dublin.

Cant, Ronald Gordon, 1970, *The University of St Andrews*, Scottish Academic Press, Edinburgh.

Caspar, Max, 1959, *Kepler* (trans. C. Doris Hellman), Dover Publications, New York (Dover edn 1993).

Chapman, Allan, 1998, *The Victorian Amateur Astronomer: Independent Astronomical Research in Britain, 1820–1920*, Wiley-Praxis, Chichester.

——2002, 'Johannes Hevelius: the last renaissance astronomer' in Patrick Moore (ed.), *2003 Yearbook of Astronomy*, Macmillan, London, pp. 246–55.

Chaucer, Geoffrey, *c.*1387, *The Canterbury Tales* (trans. by David Wright), Oxford (see Wright 1985).

Chen, P.K., 2000, 'Visions of today's giant eyes', *Sky & Telescope*, vol. 100, no. 2, pp. 34–41.

Christianson, John R., 2000, *On Tycho's Island: Tycho Brahe and His Assistants, 1570–1601*, Cambridge.

Clarke, T.N., Morrison-Low, A.D. and Simpson, A.D.C., 1989, *Brass & Glass: Scientific Instrument Making Workshops in Scotland*, National Museums of Scotland, Edinburgh.

Clay, Roger and Dawson, Bruce, 1997, *Cosmic Bullets: High Energy Particles in Astrophysics*, Allen & Unwin, Sydney.

Danjon, André and Couder, André, 1935, *Lunettes et télescopes*, Éditions de la Revue d'Optique Théorique et Instrumentale, Paris.

Davidson, D.C., 1989, *Spectacles, Lorgnettes and Monocles*, Shire Publications, Princes Risborough.

Dawe, J.A. and Watson, F.G., 1984, 'The application of optical fibre technology to Schmidt telescopes' in N. Capaccioli (ed.), *Astronomy with Schmidt-type Telescopes*, D. Reidel, Dordrecht, pp. 181–4.

Débarbat, Suzanne and Launay, Françoise, 2002, 'The objectives of the "Great Paris Exhibition Telescope" of 1900', *Bulletin of the Scientific Instrument Society*, no. 74, pp. 22–3.

Desmond, Michael and Pedretti, Carlo, 2000, *Leonardo da Vinci: The* Codex Leicester—*Notebook of a Genius*, Powerhouse Publishing, Sydney.

Dewhirst, David W. and Hoskin, Michael, 1991, 'The Rosse spirals', *Journal for the History of Astronomy*, vol. 22, pp. 257–66.

Di Cicco, Dennis, 1986, 'The journey of the 200-inch mirror', *Sky & Telescope*, vol. 71, no. 4, pp. 347–8.

Dimitroff, George Z. and Baker, James G., 1945, *Telescopes and Accessories*, Blakiston, Philadelphia.

Dobbins, Thomas and Sheehan, William, 2000, 'Beyond the Dawes limit: observing Saturn's ring divisions', *Sky & Telescope*, vol. 100, no. 5, pp. 117–21.

Dodsworth, Roger, 1982, *Glass and Glassmaking*, Shire Publications, Princes Risborough.

Dreyer, J.L.E. and Turner, H.H. (eds), 1923, *History of the Royal Astronomical Society, 1820–1920*, Royal Astronomical Society, London (rep. 1987 by Blackwell, Oxford).

ESA (European Space Agency), 1998, *The Next Generation Space Telescope: Science Drivers and Technological Challenges*, Proceedings of the 34th Liège International Astrophysics Colloquium, ESA SP-429, Noordwijk.

Frame, Tom and Faulkner, Don, 2003, *Stromlo: An Australian Observatory*, Allen & Unwin, Sydney.

Galilei, Galileo, 1610, *Sidereus Nuncius* (trans. Albert Van Helden), Chicago (see Van Helden 1989).

Gascoigne, S.C.B., 1996, 'The Great Melbourne Telescope and other 19th-century reflectors', *Quarterly Journal of the Royal Astronomical Society*, vol. 37, pp. 101–28.

Gascoigne, S.C.B., Proust, K.M. and Robins, M.O., 1990, *The Creation of the Anglo-Australian Observatory*, Cambridge.

Gilmozzi, Roberto and Dierickx, Phillipe, 2000, 'OWL concept study', *ESO Messenger*, no. 100, pp. 1–10.

Gingerich, Owen, 1979, 'The basic astronomy of Stonehenge' in Kenneth Brecher and Michael Feirtag (eds), *Astronomy of the Ancients*, MIT Press, Cambridge, Mass., pp. 117–32.

Gingrich, Mark, 2000, 'The telescope of Leonardo's dreams' (letter), *Sky & Telescope*, vol. 99, no. 3, p. 14.

Giscard d'Estaing, Valérie-Anne, 1985, *Inventions*, World Almanac Publications, New York.

Glass, I.S., 1997, *Victorian Telescope Makers: The Lives and Letters of Thomas and Howard Grubb*, Institute of Physics, Bristol.

Gleichen, Alexander, 1918, *The Theory of Modern Optical Instruments* (trans. H.H. Emsley and W. Swaine), H.M. Stationery Office, London.

Greco, Vincenzo, Molesini, Giuseppe and Quercioli, Franco, 1992, 'Optical tests of Galileo's lenses', *Nature*, vol. 358, p. 101.

Grubb Parsons, Sir Howard, & Company, 1926, *Astronomical & Optical Instruments Catalogue*, Publication No. 4, Newcastle-upon-Tyne.

——1956, *Astronomical Instruments*, Publication No. 17, Newcastle-upon-Tyne.

Hale, George Ellery, 1928, 'The possibilities of large telescopes', *Harper's Magazine*, vol. 156, pp. 639–46.

Hall, Simon (ed.), 1995, *The Hutchinson Illustrated Encyclopedia of British History*, Helicon, Oxford.

Harrison, Richard F., 1963, *Abraham Sharp, Mathematician and Astronomer, 1653–1742*, Bolling Hall Museum, Bradford.

Hart, J., van Harmelen, J., Hovey, G., Freeman, K.C., Peterson, B.A., Axelrod, T.S., Quinn, P.J., Rodgers, A.W., Allsman, R.A., Alcock, C., Bennett, D.P., Cook, K.H., Griest, K., Marshall, S.L., Pratt, M.R., Stubbs, C.W. and Sutherland, W., 1996, 'The telescope system of the MACHO program', *Publications of the Astronomical Society of the Pacific*, vol. 108, pp. 220–2.

Haynes, Raymond, Haynes, Roslynn, Malin, David and McGee, Richard, 1996, *Explorers of the Southern Sky: A History of Australian Astronomy*, Cambridge.

Hearnshaw, J.B., 1986, *The Analysis of Starlight: One Hundred and Fifty Years of Astronomical Spectroscopy*, Cambridge.

——1996, *The Measurement of Starlight: Two Centuries of Astronomical Photometry*, Cambridge.

Hetherington, Norriss S. and Brashear, Ronald S., 1992, 'Walter S. Adams and the imposed settlement between Edwin Hubble and Adriaan van Maanen', *Journal for the History of Astronomy*, vol. 23, pp. 52–6.

Hewitt-White, Ken, 2003, 'Observing Lord Rosse's spirals', *Sky & Telescope*, vol. 105, no. 5, pp. 116–21.

Hirshfeld, Alan W., 2001, *Parallax: The Race to Measure the Cosmos*, Freeman, New York.

Hodges, Paul C., 1953, 'Bernhard Schmidt and his reflector camera' in Albert G. Ingalls (ed.), *Amateur Telescope Making (Book Three)*, *Scientific American*, pp. 365–73.

Holmes, Paul, 1987, *Brahms*, Omnibus Press, London.

Hoogerdijk, Wim, et al., *c.*1994, *Binnenhof*, Information Centre Binnenhof, The Hague.

Hoskin, Michael, 1982, *Stellar Astronomy: Historical Studies*, Science History Publications, Chalfont St Giles.

——1989, 'Astronomers at war: South *v.* Sheepshanks', *Journal for the History of Astronomy*, vol. 20, pp. 175–212.

——2002, 'The Leviathan of Parsonstown: ambitions and achievements', *Journal for the History of Astronomy*, vol. 33, pp. 57–70.

——2003, 'Herschel's 40-ft reflector: funding and functions', *Journal for the History of Astronomy*, vol. 34, pp. 1–32.

Hoskin, Michael, et al., 1991, 'More on "South *v.* Sheepshanks"', *Journal for the History of Astronomy*, vol. 22, pp. 174–9.

Hoyle, Fred, 1962, *Astronomy*, Macdonald, London.

Hubble, Edwin P., 1925, 'Cepheids in spiral nebulae', *Publication of the American Astronomical Society*, vol. 5, pp. 261–4.

——1929, 'A relation between distance and radial velocity among extra-galactic nebulae', *Proceedings of the National Academy of Sciences*, vol. 15, pp. 168–73.

Humphreys, W.J., 1920, *Physics of the Air*, Franklin Institute, Philadelphia.

Hydbom, Doris, 1995, *Hven Ön i Öresund* (information leaflet), Landskrona-Vens Turistbyrå.

Károlyi, Ottó, 1965, *Introducing Music*, Penguin Books, London.

Kinder, Hermann and Hilgemann, Werner, 1974, *The Penguin Atlas of World History*, vol.1 (trans. Ernest Menze), Penguin Books, London.

King, Henry C., 1955, *The History of the Telescope*, Griffin, London.

Kizer, Kristin, 2000, 'Viking conquest of the heavens?', *Astronomy*, vol. 28, no. 9, pp. 32–4.

Koestler, Arthur, 1959, *The Sleepwalkers: A History of Man's Changing Vision of the Universe*, Hutchinson, London.

König, Albert and Köhler, Horst, 1959, *Die Fernrohre und Entfernungsmesser* (*Telescopes and Rangefinders*), 3rd edn, Springer, Berlin.

Labeyrie, Antoine, 1979, 'Standing wave and pellicle: a possible approach to very large space telescopes', *Astronomy and Astrophysics*, vol. 77, pp. L1–L2.

——1999, 'Snapshots of alien worlds: the future of interferometry', *Science*, vol. 285, pp. 1864–5.

Lacey, Robert and Danziger, Danny, 1999, *The Year 1000: What Life was Like at the Turn of the First Millennium*, Abacus, London.

Lamont-Brown, Raymond, 1989, *The Life and Times of St Andrews*, John Donald, Edinburgh.

Learner, Richard, 1986, 'The legacy of the 200-inch', *Sky & Telescope*, vol. 71, no. 4, pp. 349–53.

Leverington, David, 1995, *A History of Astronomy from 1890 to the Present*, Springer-Verlag, London.

Ley, Willy, 1963, *Watchers of the Skies*, Sidgwick & Jackson, London.

Longhurst, R.S., 1957, *Geometrical and Physical Optics*, Longmans, Green, London.

Lynch, David K. and Livingston, William, 1995, *Color and Light in Nature*, Cambridge.

Marra, Monica, 2000, 'New astronomy library in Bologna is named after Guido Horn D'Arturo: a forefather of modern telescopes', *Journal of the British Astronomical Association*, vol. 110, no. 2, p. 88.

Martin, L.C., 1948, *Technical Optics*, vol. 1, Pitman, London.

Mayr, Otto et al., 1990, *The Deutsches Museum*, Scala, London.

McConnell, Anita, 1992, *Instrument Makers to the World: A History of Cooke, Troughton & Simms*, William Sessions, York.

——1994, 'Astronomers at war: the viewpoint of Troughton & Simms', *Journal for the History of Astronomy*, vol. 25, pp. 219–35.

McCrea, W.H., 1975, *The Royal Greenwich Observatory: An Historical Review issued on the Occasion of its Tercentenary*, H.M. Stationery Office, London.

McKenna-Lawlor, Susan and Hoskin, Michael, 1984, 'Correspondence of Markree Observatory', *Journal for the History of Astronomy*, vol. 15, pp. 64–8.

Mills, A. (attrib.), 1992, 'Did an Englishman invent the telescope? Leonard Digges' "Perspective" of 1560', *Bulletin of the Scientific Instrument Society*, no. 35, p. 2.

——1993, 'Postscript—nature got there first!', *Bulletin of the Scientific Instrument Society*, no. 37, p. 10.

Misch, Tony and Stone, Remington, *James Lick, the 'Generous Miser': The Building of Lick Observatory*, University of California Observatories/Lick Observatory, http://www.ucolick.org/ [February 2002]

Molesini, Giuseppe, 2003, 'The telescopes of seventeenth-century Italy', *Optics & Photonics News*, vol. 14, no. 6, pp. 34–9.

Møller, Palle (ed.), 1996, *Hubble Space Telescope Cycle 7 Call for Proposals*, Space Telescope Science Institute, Baltimore.

Moore, Patrick, 1996, *The Planet Neptune: An Historical Survey Before Voyager*, 2nd edn, John Wiley, Chichester.

——1997, *Eyes on the Universe: The Story of the Telescope*, Springer, London.

Moss, Michael and Russell, Iain, 1988, *Range and Vision: The First 100 Years of Barr and Stroud*, Mainstream Publishing, Edinburgh.

Mountain, Matt, and Gillett, Fred, 1998, 'The revolution in telescope aperture', *Nature*, supplement to vol. 395, no. 6701, pp. A23–A29.

Murdin, Paul, 1998, 'The origin of cosmic gamma-ray bursts' in Patrick Moore (ed.), *1999 Yearbook of Astronomy*, Macmillan, London, pp. 169–79.

Natarajan, Priyamvada, 1998, 'The Universe through gravity's lens' in Peter Coles (ed.), *The Icon Critical Dictionary of the New Cosmology*, Icon Books, Cambridge, pp. 99–114.

Newton, Isaac, 1730, *Opticks*, 4th edn, Dover Publications, New York (Dover edn 1979).

Nicholl, Charles, 2001, 'A "mad priest of the sun" burns', *BBC History Magazine*, vol. 2, no. 2, pp. 44–5.

Nicolson, Iain, 2001, 'A Universe of darkness' in Sir Patrick Moore (ed.), *2002 Yearbook of Astronomy*, Macmillan, London, pp. 243–64.

Osterbrock, Donald E., 1994, 'Getting the picture: wide-field astronomical photography from Barnard to the achromatic Schmidt, 1888–1992', *Journal for the History of Astronomy*, vol. 25, pp. 1–14.

O'Connor, John J. and Robertson, Edmund F., *The MacTutor History of Mathematics Archive*, http://www-history.mcs.st-andrews.ac.uk/history/index.html [May 2004]

Panek, Richard, 2000, *Seeing and Believing: The Story of the Telescope, or How We Found Our Place in the Universe*, Fourth Estate, London.

Proust, Dominique, 1992, 'William Herschel (1738–1822)—organ works', sleeve note for *Pièces d'orgue de William Herschel*, Disques Dom, Vincennes (DOM CD 1418).

Reid, William, 1983, 'Binoculars in the Army, Part II, 1904–19' in Elizabeth Talbot Rice and Alan Guy (eds), *Army Museum '82*, National Army Museum, London, pp. 15–30.

——2001, *'We're Certainly Not Afraid of Zeiss': Barr & Stroud Binoculars and the Royal Navy*, National Museums of Scotland, Edinburgh.

Riekher, Rolf, 1990, *Fernrohre und ihre Meister (Telescopes and Their Masters)*, 2nd edn, Verlag Technik GmbH, Berlin.

Ronan, Colin A., 1991, 'The origins of the reflecting telescope', *Journal of the British Astronomical Association*, vol. 101, no. 6, pp. 335–42.

Ronan, Colin A., Turner, G.L'E., Darius, J., Rienitz, J., Howse, D. and Ringwood, S.D., 1993, 'Was there an Elizabethan telescope?', *Bulletin of the Scientific Instrument Society*, no. 37, pp. 2–10.

Roslund, Curt, 1989, 'Tycho Brahe's innovations in instrument design', *Bulletin of the Scientific Instrument Society*, no. 22, pp. 2–4.

Russell, H.C. (attrib.), 1892, *Description of the Star Camera, at the Sydney Observatory*, Minister for Public Instruction, Sydney.

Schilling, Govert, 2000, 'Giant eyes of the future', *Sky & Telescope*, vol. 100, no. 2, pp. 52–6.

Schmidt, Bernhard, 1931, 'Ein lichtstarkes Komafreies Spiegelsystem', *Zentralzeitung für Optik und Mechanik*, vol. 52, pp. 25–6. (Trans. Nicholas U. Mayall, 1946, 'A rapid coma-free mirror system', *Publications of the Astronomical Society of the Pacific*, vol. 58, pp. 285–90.)

Schmidt, Erik, 1995, *Optical illusions: The Life Story of Bernhard Schmidt, the Great Stellar Optician of the Twentieth Century*, Estonian Academy Publishers.

Schütz, Wilhelm, 1966, 'Ernst Abbe: university teacher and industrial physicist' in *Carl Zeiss: 150th Anniversary of his Birthday* (supplement to *Jena Review*), pp. 13–23, Carl Zeiss, Jena.

Simpson, A.D.C., 1985, 'Richard Reeve—the "English Campani"—and the origins of the London telescope-making tradition', *Vistas in Astronomy*, vol. 28, pp. 357–65.

——1992, 'James Gregory and the reflecting telescope', *Journal for the History of Astronomy*, vol. 23, pp. 77–92.

Sisson, George, 1989, 'David Scatcherd Brown (1927–1987)', *Quarterly Journal of the Royal Astronomical Society*, vol. 30, pp. 279–81.

Sluiter, Engel, 1997a, 'The first known telescopes carried to America, Asia and the Arctic, 1614–39', *Journal for the History of Astronomy*, vol. 28, pp. 141–5.

——1997b, 'The telescope before Galileo', *Journal for the History of Astronomy*, vol. 28, pp. 223–34.

Sobel, Dava, 1999, *Galileo's Daughter*, Fourth Estate, London.

Spaight, John Tracy, 2004, '"For the good of astronomy": the manufacture, sale and distant use of William Herschel's telescopes', *Journal for the History of Astronomy*, vol. 35, pp. 45–69.

Spargo, P.E., 1984, 'Burning glasses', *Bulletin of the Scientific Instrument Society*, no. 4, pp. 7–8. (See also the erratum in *Bulletin of the Scientific Instrument Society*, no. 5, p. 23.)

SPIE, 2000a, *Telescope Structures, Enclosures, Controls, Assembly/Integration/Validation and Commissioning, Proc. SPIE*, vol. 4004.

——2000b, *Discoveries and Research Prospects from 8–10 Meter-Class Telescopes, Proc. SPIE*, vol. 4005.

——2000c, *Adaptive Optical Systems Technology, Proc. SPIE*, vol. 4007.

——2000d, *Optical and IR Telescope Instrumentation and Detectors, Proc. SPIE*, vol. 4008.

Steel, Duncan, 2002, 'Near-Earth objects: getting up close and personal' in Sir Patrick Moore (ed.), *2003 Yearbook of Astronomy*, Macmillan, London, pp. 154–80.

Stooke, Philip, 1996, 'The mirror in the Moon', *Sky & Telescope*, vol. 91, no. 3, pp. 96–8.

Strandh, Sigvard, 1979, *Machines: An Illustrated History*, AB Nordbok, Gothenburg.

Talbot, Stuart, 1996, 'Jesse Ramsden F.R.S.: his optical testament', *Bulletin of the Scientific Instrument Society*, no. 50, pp. 27–9.

——2002, 'The astroscope by James Mann of London: the first commercial achromatic refracting telescope *c.*1735', *Bulletin of the Scientific Instrument Society*, no. 75, pp. 6–8.

Temple, Robert, 2000, *The Crystal Sun: Rediscovering a Lost Technology of the Ancient World*, Century, London.

Thackeray, A.D., 1961, *Astronomical Spectroscopy*, Eyre & Spottiswoode, London.

Thiel, Rudolf, 1958, *And There was Light: The Discovery of the Universe* (trans. Richard and Clara Winston), Andre Deutsch, London.

Thoday, A.G., 1971, *Astronomy 2: Astronomical Telescopes*, Science Museum, H.M. Stationery Office, London.

Thoren, Victor E. (with contributions by John R. Christianson), 1990, *The Lord of Uraniborg: A Biography of Tycho Brahe*, Cambridge.

Tomalin, Claire, 2002, *Samuel Pepys: The Unequalled Self*, Penguin Books, London.

Turnbull, H.W. (ed.), 1959, *The Correspondence of Isaac Newton, Vol. 1: 1661–1675*, Cambridge.

Turner, A.J., 1977, 'Some comments by Caroline Herschel on the use of the 40-ft telescope', *Journal for the History of Astronomy*, vol. 8, pp. 196–8.

——2002, 'The observatory and the quadrant in eighteenth-century Europe', *Journal for the History of Astronomy*, vol. 33, pp. 373–85.

Van Helden, Albert, 1977a, 'The development of compound eyepieces, 1640–1670', *Journal for the History of Astronomy*, vol. 8, pp. 26–37.

——1977b, 'The invention of the telescope', *Transactions of the American Philosophical Society*, vol. 67, part 4.

——1989, 'Introduction' and 'Conclusion' to the translation of Galileo's *Sidereus Nuncius*, Chicago.

Wachtmeister, Hélène and Wachtmeister, Henrik, c.1996, *Welcome to Knutstorp*, privately-produced leaflet.

Wailes, Rex, 1963, 'James Nasmyth—artist's son' in The Institute of Mechanical Engineers, *Engineering Heritage: Highlights from the History of Mechanical Engineering*, Heinemann, London, pp. 106–11.

Waller, Maureen, 2000, *1700: Scenes from London Life*, Hodder & Stoughton, London.

Warner, Brian, 1975, 'A forgotten 41-inch refractor', *Sky & Telescope*, vol. 50, no. 6, p. 370.

——1982, 'The Large Southern Telescope: Cape or Melbourne?', *Quarterly Journal of the Royal Astronomical Society*, vol. 23, pp. 505–14.

Watson, Fred, 1995, *Binoculars, Opera Glasses and Field Glasses*, Shire Publications, Princes Risborough.

——1999a, 'How Zeiss binoculars made their London début', *Zeiss Historica*, vol. 21, no. 2, pp. 4–11.

——1999b, 'Optical astronomy, the early Universe and the telescope superleague' in Patrick Moore (ed.), *2000 Yearbook of Astronomy*, Macmillan, London, pp. 178–204.

——2000, 'The dawn of binocular astronomy' in Patrick Moore (ed.), *2001 Yearbook of Astronomy*, Macmillan, London, pp. 162–83.

——2001, 'The enduring legacy of Bernhard Schmidt' in Sir Patrick Moore (ed.), *2002 Yearbook of Astronomy*, Macmillan, London, pp. 224–42.

——2002, 'Newton's telescope and the half-filled bathtub', *Anglo-Australian Observatory Newsletter*, no. 100, pp. 14–15.

——2003a, 'Absolutely nebulous' in Sir Patrick Moore (ed.), *2004 Yearbook of Astronomy*, Macmillan, London, pp. 233–44.

——2003b, 'Optical spectroscopy today and tomorrow' in John Mason (ed.), *Astrophysics Update*, Springer-Praxis, Chichester, pp. 185–214.

Watson, W., & Sons Ltd, *c*.1925, *A Catalogue of Binoculars and Telescopes* (44th edn), London.

Welther, Barbara L., 1999, 'Leonardo da Vinci and the Moon', *Sky & Telescope*, vol. 98, no. 4, pp. 40–4.

Wennberg, Arne, 1996, *Tänk, om det är så! Om Tycho Brahes instrument och vad han kunde göra med dessa*, Maxi Data HB, Landskrona.

Wesley, Walter G., 1978, 'The accuracy of Tycho Brahe's instruments', *Journal for the History of Astronomy*, vol. 9, pp. 42–53.

West, Richard M., 1997, 'Tycho and his observatory as sources of inspiration to modern astronomy' in Arne Ardeberg (ed.), *Optical Telescopes of Today and Tomorrow: Following in the Direction of Tycho Brahe, Proc. SPIE*, vol. 2841, pp. 774–83.

Westfall, Richard S., 1989, 'The trial of Galileo: Bellarmino, Galileo and the clash of two worlds', *Journal for the History of Astronomy*, vol. 20, pp. 1–23.

——1991, 'Victor E. Thoren (1935–1991)', *Journal for the History of Astronomy*, vol. 22, pp. 253–4.

White, Michael, 2000, *Leonardo: The First Scientist*, Little Brown, London.

Willach, Rolf, 2001, 'The development of lens grinding and polishing techniques in the first half of the 17th century', *Bulletin of the Scientific Instrument Society*, no. 68, pp. 10–15.

——2002, 'The Wiesel telescopes in Skokloster Castle and their historical background', *Bulletin of the Scientific Instrument Society*, no. 73, pp. 17–22.

Wilson, R.N., 1996, *Reflecting Telescope Optics I*, Springer, Berlin.

——1999, *Reflecting Telescope Optics II*, Springer, Berlin.

Wright, David, 1985, 'Introduction' to his translation of Chaucer's *The Canterbury Tales*, Oxford.

Yapp, Nick, 2000, *The British Millennium: 1000 Remarkable Years of Incident and Achievement*, Könemann, Köln.

Zeiss, Carl, GmbH, 1996, *Anticipating the Future*, Microscopes Business Unit, Carl Zeiss, Jena.

GLOSSARY

absorption lines the dark lines appearing in a star's **spectrum.** They are caused by the gases in the star's atmosphere absorbing light at particular wavelengths (colours), and provide detailed information about the star.

achromatic lens a compound lens in which **chromatic aberration** is corrected by the use of multiple component lenses. The most common achromatic lens is a **doublet** (i.e., two component lenses).

adaptive optics telescope **instrumentation** that attempts to correct images blurred by poor **seeing** to restore lost detail.

altazimuth a **mounting** for a telescope that allows it to be moved about a vertical (**azimuth**) axis and a horizontal (**altitude**) axis. It acts in the same way as a surveyor's theodolite, allowing the telescope to be pointed to any part of the sky. Unlike the **equatorial**, movements of both axes are needed to track the stars.

altitude the angular height of a celestial object above the horizon (ranging from zero degrees at the horizon to 90 degrees at the **zenith**).

aperture the diameter of the light-collecting component of a telescope, usually its mirror (for a **reflector**) or lens (for a **refractor**).

apochromat a compound lens, usually in the form of a **triplet**, with a very high degree of colour correction.

arcsecond (or second of arc) an angle equal to 1/3600th of a degree.

aspheric shape of an optical surface that is not a segment of a sphere, e.g. a **paraboloid** or an **ellipsoid**.

astatic a type of mirror support that uses pivoted counterweights to balance the load.

astronomical telescope formerly an alternative name for a **Keplerian telescope**, but today means any telescope used for astronomy.

atmospheric turbulence chaotic movement in a flow of air.

azimuth the compass-bearing of a celestial object, measured in degrees from north (through east).

binocular (as an adjective) two-eyed; (as a noun) a nineteenth-century contraction of 'binocular field glass'. It is now usually used in the plural to mean the prismatic type.

black hole an object of such extreme density that nothing can escape its gravitational pull, not even light.

burning glass a lens or mirror (used singly or in combination) for heating or igniting material using sunlight.

Cassegrain telescope a **reflecting telescope** design published by Cassegrain (but probably invented by Gregory) using a concave **paraboloidal** mirror in combination with a **hyperboloidal** convex secondary mirror and an eyepiece.

catadioptric using or concerned with both the reflection and refraction of light.

catoptric using or concerned with the reflection of light.

celestial pole the point in the sky about which the whole celestial sphere appears to rotate. Its altitude is equal to the observer's latitude.

celestial sphere an imaginary sphere centred on the observer on which all objects in the sky seem to be projected.

chromatic aberration coloured halos in the images formed by a lens resulting from the unequal refraction of different colours at its surfaces.

coma an error in the image of a star away from the centre of the field of view causing it to take on the appearance of a comet with a short tail.

concave lens one that is thinner in the centre than the edge.

convex lens one that is thicker in the centre than the edge.

crown glass common type of glass named after the blow-and-spin method by which it was originally produced.

declination the equivalent of terrestrial latitude on the celestial sphere; i.e. angular distance from the equator.

diffraction grating a glass plate or mirror with fine lines ruled on it that acts in a similar way to a prism. It disperses white light into a spectrum by the process of diffraction.

dioptric using or concerned with the refraction of light.

doublet an achromatic lens consisting of two components close together or in contact.

drawtube telescope usually a long terrestrial telescope that can be collapsed using a series of tubes to make it more portable.

Dutch telescope an archaic term for a Galilean telescope.

electromagnetic radiation energy consisting of electric and magnetic fields vibrating at right angles to one another, travelling through space at the speed of light.

ellipsoid the shape formed by rotating the closed curve known as an ellipse about its long axis. A concave mirror made as a segment of this shape is used in combination with a paraboloid to form a Gregorian telescope.

English equatorial a form of equatorial in which the polar axis is supported at both ends and the declination axis takes the form of a crosspiece in the middle. It is an older form than the German equatorial.

equatorial a mounting for a telescope that allows it to be moved about an axis parallel to the Earth's (called the polar axis, because it points to the celestial pole). A second axis at right angles to this (the declination axis) allows the telescope to be pointed to any part of the sky. Once an object has been found, the telescope only needs to be moved about the polar axis to track it.

erecting lens a convex lens (or group of lenses) placed between the objective and eyepiece of a Keplerian telescope to turn the image upright.

eye lens the lens nearest the eye in a multi-element eyepiece.

eyepiece a lens or combination of lenses close to the eye that magnifies the image of a distant scene formed by the objective.

field lens a lens designed to increase the field of view of a multi-element eyepiece.

field of view the angular diameter of the circular area seen through an optical instrument, usually quoted in degrees.

flint glass a dense, brilliant glass with significantly different optical properties from ordinary crown glass.

focal length the distance from a lens or mirror to its focal point.

focal point the position at which a lens or mirror forms an image of a distant object.

Fraunhofer lines the absorption lines in the spectrum of the Sun.

galaxy a gigantic system of stars, gas and dust, exemplified by the Milky Way Galaxy of which the Sun is a member. Galaxies, now known in very large numbers throughout the Universe, were originally indistinguishable from nebulae

Galilean telescope a simple combination of a convex and concave lens requiring no additional optical components to make the magnified images of distant objects upright.

gamma rays type of electromagnetic radiation with a wavelength less than 0.04 nanometres.

German equatorial a form of equatorial in which the polar and declination axes are in the form of a T, with the telescope on one side and a counterweight on the other.

geodesy the study of the shape and size of the Earth.

globular cluster a spherical or near-spherical aggregation of very old stars, now known to exist in large numbers around galaxies (including our own).

gravitational lens the distortion of space produced by a massive object such as a galaxy or cluster of galaxies. It acts like a crude natural lens of vast proportions.

Gregorian telescope Gregory's design for a reflecting telescope design which used a concave paraboloidal mirror in combination with an ellipsoidal concave secondary mirror and an eyepiece.

Huygens eyepiece a particular combination of two convex lenses (a field lens and an eye lens) that improves field of view and image quality in an optical instrument. Its focal point is between the two lenses.

hyperboloid the shape formed by rotating the open curve known as a hyperbola about its axis. A convex mirror made as a segment of this shape is used in combination with a paraboloid to form a Cassegrain telescope.

infrared type of electromagnetic radiation with a wavelength between 1000 nanometres and 0.35 millimetres.

instrumentation auxiliary equipment attached to a telescope which measures and analyses the light or radiation from celestial objects.

Keplerian, astronomical or inverting telescope a combination of two convex lenses (or groups of lenses) that produces magnified inverted images of distant objects.

light-year the distance light travels in one year, equal to 9.5 million million km.

magnification the number of times larger a distant object appears viewed through a telescope rather than the unaided eye.

micrometer (or **eyepiece micrometer**) a device with moveable threads used to measure small angles using a telescope (e.g. the diameter of a planet or the separation of a double star).

millimetre waves (microwaves) type of **electromagnetic radiation** with a wavelength between 1 and 30 millimetres.

mirror cell the mechanical structure supporting a telescope mirror.

monolithic mirror a telescope mirror that is made from one piece of glass.

mounting (of a telescope) the structure that supports the **tube** of the telescope and allows it to be pointed in different directions.

nanometre (nm) a millionth of a millimetre.

nebula a celestial object appearing indistinct and misty. Originally used for all such objects, the term now refers specifically to clouds of gas and dust.

neutron star an extremely dense star a few km in diameter, which is prevented from collapsing to a **black hole** only by the pressure of neutrons.

Newtonian telescope Newton's **reflecting telescope** design which used a concave **paraboloidal** mirror in combination with a flat **secondary mirror** and an **eyepiece**.

objective a lens or mirror that is the main light collecting component of a telescope and forms images of distant objects.

ocular the archaic term for an **eyepiece**.

optical aberration a defect in the image produced by a lens or mirror system, causing blurring or spurious colour. Examples are **spherical aberration, chromatic aberration** and **coma**.

optical telescope one designed to detect visible light.

orrery a small mechanical model of the Solar System in which the relative motions of the planets are mimicked by gears. Named after Charles Boyle, fourth Earl of Cork and Orrery.

paraboloid the shape of a mirror surface that can form an image of a distant object free from **spherical aberration**. It is generated mathematically by rotating the open curve known as a parabola about its axis.

primary mirror the main light-collecting mirror of a **reflecting telescope**.

quasar the highly luminous core of a young galaxy powered by a supermassive **black hole**.

radio waves type of **electromagnetic radiation** with wavelengths longer than about 10 millimetres.

Ramsden eyepiece a particular combination of two convex lenses (a **field lens** and an **eye lens**) that improves field of view and image quality in an optical instrument. Its **focal point** is outside the two lenses.

reflecting telescope or reflector a telescope whose main light collecting component is a curved (**concave**) mirror.

refracting telescope or refractor a telescope whose main light-collecting component is a **convex glass lens** or combination of lenses.

refraction the bending of light as it crosses from one transparent medium to another (e.g. from air into glass).

resolution the fineness of the detail that can be revealed by a telescope.

secondary mirror a mirror used to deflect or refocus the light from the **primary mirror** in a **reflecting telescope**.

seeing the diameter of a star image blurred by **atmospheric turbulence**, measured in **arcseconds**.

segmented mirror a telescope mirror that is made up of many close-fitting hexagonal segments, each independently computer-controlled in alignment so that they act together as a single mirror.

sighting tube an empty tube resembling a telescope, used to reduce glare or define a line of sight.

spectrograph the instrument used to record the **spectrum** of a celestial object.

spectrum (pl. spectra) the light from a celestial object or other light-source spread into the rainbow of colours that represent different wavelengths. It reveals the physical information imprinted into the light.

speculum the archaic term for the mirror of a **reflecting telescope**.

speculum metal an alloy of copper, tin and other materials formerly used to make mirrors.

spherical aberration the blurring of the image formed by a mirror or lens resulting from the spherical curvature of its surface(s).

spyglass an archaic term that originally meant a small **Galilean telescope**. Common usage now includes **terrestrial telescopes**.

star catalogue a detailed list of the positions and brightness of stars.

stellar spectroscopy the study of the **spectra** of stars.

supernova the cataclysmic explosion of a massive star at the end of its life.

terrestrial telescope an instrument giving an upright image. The term usually means a **Keplerian telescope** with the addition of an **erecting lens**, rather than a simple **Galilean telescope**.

triplet a compound lens consisting of three components close together or in contact, usually used to improve colour correction.

tube (of a telescope) originally, the enclosed tube that held together the optical components of the telescope and prevented the entry of extraneous light. In modern telescopes it refers to the open structure that supports the optical components.

ultraviolet rays type of **electromagnetic radiation** with wavelengths between about 10 and 350 **nanometres**.

Universe everything that can ever be detected, encompassing all of space, time, matter and radiation.

visible light type of **electromagnetic radiation** with wavelengths conventionally taken to be between about 350 and 1000 **nanometres** (rather wider than the sensitivity of the eye).

wide-field or wide-angle having a large **field of view**.

X-rays type of **electromagnetic radiation** with wavelengths between about 0.04 and 10 **nanometres**.

zenith the point directly overhead.

THE WORLD'S GREAT TELESCOPES

A list of the largest ground based optical (visible light) and infrared telescopes operating or nearing operation in 2004.

REFLECTING TELESCOPES WITH APERTURES OF 3.6 METRES OR GREATER.

Very Large Telescope (VLT) Four 8.2 metre (16.4 metre equivalent when combined) thin monolithic mirrors. *Location:* Cerro Paranal, Chile, 2635 metres. Completed 1998–2001, and named Antu, Kueyen, Melipal and Yepun. *Operated by:* European Southern Observatory.

Keck Telescopes (I and II) Two 9.8 metre (13.9 metre equivalent when combined) segmented mirrors. *Location:* Mauna Kea, Hawaii, 4145 metres. Completed 1991 and 1996. *Operated by:* W.M. Keck Observatory (University of California, Caltech and NASA).

Large Binocular Telescope (LBT) Two 8.4 metre (11.9 metre equivalent) spin-cast monolithic mirrors. *Location:* Mt Graham, Arizona, 3260 metres. Operational 2005. *Operated by:* University of Arizona, Osservatorio Astrofisico de Arcetri (Italy) and other US and German partners.

Gran Telescopio Canarias (GTC) 10.4 metre segmented mirror. *Location:* Roque de los Muchachos, La Palma, Canary Islands, 2400 metres. Scheduled 2004. *Operated by:* Instituto de Astrofísica de Canarias, Spain with other partners.

Hobby-Eberly Telescope (HET) 9.1 metre (effective aperture) segmented mirror. *Location:* Mt Fowlkes, Texas, 2025 metres. Completed 1997. *Operated by:* consortium of US and German universities.

Southern African Large Telescope (SALT) 9.1 metre (effective aperture) segmented mirror. *Location:* Sutherland, South Africa, 1798 metres. Scheduled 2004. *Operated by:* consortium of South Africa, Poland, New Zealand and US universities.

Subaru 8.2 metre thin monolithic mirror. *Location:* Mauna Kea, Hawaii, 4139 metres. Completed 1999. *Operated by:* National Astronomical Observatory of Japan.

Gemini North Telescope 8.1 metre thin monolithic mirror. *Location:* Mauna Kea, Hawaii, 4214 metres. Completed 1999. *Operated by:* Gemini Observatory, a consortium of USA, UK, Canada, Australia, Argentina, Brazil and Chile.

Gemini South Telescope 8.1 metre thin monolithic mirror. *Location:* Cerro Pachón, Chile, 2715 metres. Completed 2002. *Operated by:* Gemini Observatory, a consortium of USA, UK, Canada, Australia, Argentina, Brazil and Chile.

MMT Observatory 6.5 metre spin-cast monolithic mirror. *Location:* Mt Hopkins, Arizona, 2606 metres. Completed 2000 (modified from the Multiple-Mirror Telescope, built in 1979). *Operated by:* Smithsonian Institution and University of Arizona.

Magellan (I and II) Two 6.5 metre spin-cast monolithic mirrors. *Location:* Las Campanas, Chile, 2300 metres. Completed 2000 and 2002, and named the Baade and Clay telescopes. *Operated by:* Carnegie Institution, Harvard, Michigan, Arizona and MIT.

Bolshoi Telescope Azimutal'ny (BTA) Massive 6.0 metre monolithic mirror. *Location:* Mt Pastukhov, Russia, 2100 metres. Completed 1976. *Operated by:* Special Astrophysical Observatory, Russian Academy of Sciences.

Large Zenith Telescope (LZT) 6.0 metre liquid mirror formed from a rotating dish of mercury. *Location:* Maple Ridge, BC, Canada, 395 metres. Completed 2001. *Operated by:* University of British Colombia, Laval University and Institute d'Astrophysique de Paris.

Hale Telescope Massive 5.1 metre monolithic mirror. *Location:* Palomar Mountain, California, 1706 metres. Completed 1948. *Operated by:* Caltech.

William Herschel Telescope Massive 4.2 metre monolithic mirror. *Location:* Roque de los Muchachos, La Palma, Canary Islands, 2332 metres. Completed 1987. *Operated by:* Isaac Newton Group of Telescopes, a consortium of the UK, the Netherlands and Spain.

SOAR Telescope 4.2 metre thin monolithic mirror. *Location:* Cerro Pachón, Chile, 2701 metres. Completed 2002. *Operated by:* Southern Observatory for Astrophysical Research, a consortium of Brazil, the US National Optical Astronomy Observatory and US universities.

Victor M. Blanco Telescope Massive 4.0 metre monolithic mirror. *Location:* Cerro Tololo, Chile, 2215 metres. Completed 1976. *Operated by:* Cerro Tololo Inter-American Observatory, funded by the US National Science Foundation and Association of Universities for Research in Astronomy.

Nicholas U. Mayall Telescope Massive 4.0 metre monolithic mirror. *Location:* Kitt Peak, Arizona, 2120 metres. Completed 1973. *Operated by:* Kitt Peak National Observatory.

Anglo-Australian Telescope Massive 3.9 metre monolithic mirror. *Location:* Siding Spring Mountain, Australia, 1150 metres. Completed 1974. *Operated by:* Anglo-Australian Observatory.

United Kingdom Infrared Telescope 3.8 metre monolithic mirror. *Location:* Mauna Kea, Hawaii, 4194 metres. Completed 1979. *Operated by:* Joint Astronomy Center, a consortium of the UK, Canada and the Netherlands.

Advanced Electro-Optical System Telescope 3.6 metre thin monolithic mirror. *Location:* Haleakala, Hawaii, 3058 metres. Completed 2000. *Operated by:* US Air Force Research Laboratory.

Canada-France-Hawaii Telescope Massive 3.6 metre monolithic mirror. *Location:* Mauna Kea, Hawaii, 4200 metres. Completed 1979. *Operated by:* Canada-France-Hawaii Telescope Corporation.

Telescopio Nazionale Galileo 3.6 metre thin monolithic mirror. *Location:* Roque de los Muchachos, La Palma, Canary Islands, 2370 metres. Completed 1997. *Operated by:* Centro Galileo Galilei for the Consorzio Nazionale per l'Astronomia e l'Astrofisica (Italy).

ESO 3.6-m Telescope Massive 3.6 metre monolithic mirror. *Location:* La Silla, Chile, 2387 metres. Completed 1976. *Operated by:* European Southern Observatory.

REFRACTING TELESCOPES WITH APERTURES OF 70 CM OR GREATER.

Yerkes 40-inch Refractor 102 cm lens. *Location:* Williams Bay, Wisconsin, 334 metres. Completed 1897. *Operated by:* University of Chicago.

Lick 36-inch Refractor 91 cm lens. *Location:* Mt Hamilton, California, 1280 metres. Completed 1888. *Operated by:* Lick Observatory, University of California.

Meudon 33-inch (Grande Lunette) 83 cm lens. *Location:* Meudon, France, 162 metres. Completed 1889. *Operated by:* Observatoire de Paris.

Potsdam Refractor 80 cm lens. *Location:* Potsdam, Germany, 107 metres. Completed 1899. *Operated by:* Astrophysikalisches Institut Potsdam.

Thaw Refractor 76 cm lens. *Location:* Pittsburgh, Philadelphia, 380 metres. Completed 1912. *Operated by:* Allegheny Observatory, University of Pittsburgh.

Lunette Bischoffscheim 74 cm lens. *Location:* Mont Gros, France, 372 metres. Completed 1886. *Operated by:* Observatoire de la Côte d'Azur.

Greenwich 28-inch Refractor   71 cm lens. *Location:* Greenwich, UK, 47 metres. Completed 1893. *Operated by:* Royal Observatory Greenwich.

SCHMIDT TELESCOPES WITH APERTURES OF 1.00 METRES OR GREATER.

LAMOST (Large-Area Multi-Object Survey Telescope) 5.70 × 4.40 metre segmented reflective corrector plate feeding a 6.70 × 6.00 metre segmented spherical mirror. *Location:* Xinglong, China, 960 metres. Scheduled 2006. *Operated by:* Beijing Astronomical Observatory.

Tautenburg Schmidt Telescope   1.34 metre corrector plate feeding a 2.00 metre spherical mirror. *Location:* Tautenburg, Germany, 331 metres. Completed 1960. *Operated by:* Karl Schwarzschild Observatorium.

Oschin (Palomar) Schmidt Telescope   1.24 metre achromatic corrector plate feeding a 1.83 metre spherical mirror. *Location:* Palomar Mountain, California, 1706 metres. Completed 1948. *Operated by:* Caltech.

United Kingdom Schmidt Telescope   1.24 metre achromatic corrector plate feeding a 1.83 metre spherical mirror. *Location:* Siding Spring Mountain, Australia, 1145 metres. Completed 1973. *Operated by:* Anglo-Australian Observatory.

Kiso Schmidt Telescope   1.05 metre corrector plate feeding a 1.50 metre spherical mirror. *Location:* Kiso, Japan, 1130 metres. Completed 1976. *Operated by:* University of Tokyo.

ESO Schmidt Telescope   1.00 metre achromatic corrector plate feeding a 1.62 metre spherical mirror. *Location:* La Silla, Chile, 2318 metres. Completed 1972. *Operated by:* European Southern Observatory.

Llano del Hato Schmidt Telescope   1.00 metre corrector plate feeding a 1.52 metre spherical mirror. *Location:* Mérida, Venezuela, 3610 metres. Completed 1978. *Operated by:* Centro F.J. Duarte, Venezuela.

Byurakan Schmidt Telescope 1.00 metre corrector plate feeding a 1.50 metre spherical mirror. *Location:* Mt Aragatz, Armenia, 1450 metres. Completed 1961. *Operated by:* Byurakan Astrophysical Observatory.

Kvistaberg Schmidt Telescope   1.00 metre corrector plate feeding a 1.35 metre spherical mirror. *Location:* Kvistaberg, Sweden, 33 metres. Completed 1963. *Operated by:* Uppsala Astronomical Observatory, Uppsala University.

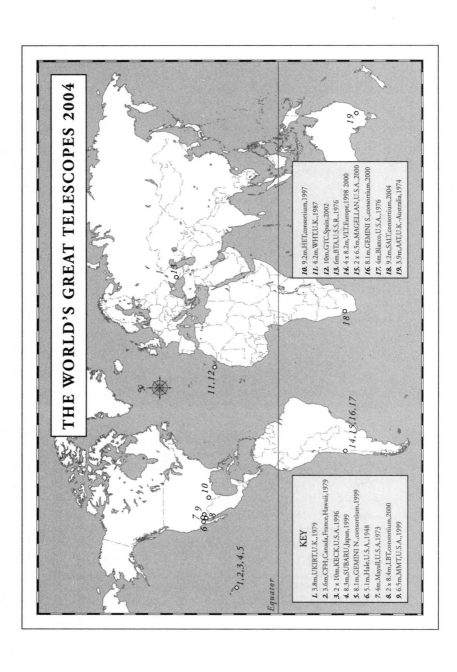

THE WORLD'S GREAT TELESCOPES 2004

KEY

1. 3.8m,UKIRT,U.K.,1979
2. 3.6m,CFH,Canada,France,Hawaii,1979
3. 2 x 10m,KECK,U.S.A.,1996
4. 8.3m,SUBARU,Japan,1999
5. 8.1m,GEMINI N.,consortium,1999
6. 5.1m,Hale,U.S.A.,1948
7. 4m,Mayall,U.S.A,1973
8. 2 x 8.4m,LBT,consortium,2000
9. 6.5m,MMT,U.S.A.,1999

10. 9.2m,HET,consortium,1997
11. 4.2m,WHT,U.K.,1987
12. 10m,GTC,Spain,2002
13. 6m,BTA,U.S.S.R.,1976
14. 4 x 8.2m,VLT,Europe,1998 2000
15. 2 x 6.5m,MAGELLAN,U.S.A.,2000
16. 8.1m,GEMINI S.,consortium,2000
17. 4m,Blanco,U.S.A.,1976
18. 9.2m,SALT,consortium,2004
19. 3.9m,AAT,U.K.-Australia,1974

Equator

INDEX